Coenzyme Contents of Arterial Tissue

Monographs on Atherosclerosis

Vol. 4

Editors
JOHN E. KIRK, St. Louis, Mo.
DAVID KRITCHEVSKY, Philadelphia, Pa.
O. J. POLLAK, Dover, Del.
HENRY S. SIMMS, Rockleigh, N. J.

S. Karger · Basel · München · Paris · London · New York · Sydney

Coenzyme Contents of Arterial Tissue

J. E. KIRK
Washington University School of Medicine, St. Louis, USA

With 44 tables

S. Karger · Basel · München · Paris · London · New York · Sydney

Monographs on Atherosclerosis

Vol. 1: O. J. POLLAK (Dover, Del.):
Tissue Cultures. XII+143 pp., 28 fig. comprising 49 black and white and
32 color illustrations, 9 tab., 1969
ISBN 3-8055-0437-3

Vol. 2: J. C. GEER (Columbus, Ohio) and DARIA M. HAUST (London, Ontario):
Smooth Muscle Cells in Atherosclerosis. X+140 pp., 27 fig., 1972
ISBN 3-8055-1377-1

Vol. 3: J. E. KIRK (St. Louis, Mo.):
Vitamin Contents of Arterial Tissue. XII+104 pp., 62 tab., 1973
ISBN 3-8055-1466-2

S. Karger · Basel · München · Paris · London · New York · Sydney
Arnold-Böcklin-Strasse 25, CH-4011 Basel (Switzerland)

All rights, including that of translation into other languages, reserved.
Photomechanic reproduction (photocopy, microcopy) of this book or part of it without
special permission of the publishers is prohibited.

© Copyright 1974 by S. Karger AG, Verlag für Medizin und Naturwissenschaften, Basel
Printed in Switzerland by Buchdruckerei Gasser & Eggerling AG, Chur
Blocks: Steiner & Co., Basel
ISBN 3-8055-1670-3

Dedicated to my wife

Contents

	Acknowledgements	VIII
	List of Abbreviations	IX
I.	Introduction	1
	References	2
II.	Effects of some Coenzymes and Metabolic Factors on Experimental Atherosclerosis in Animals	3
	References	5
III.	Carnitine (Free Carnitine)	6
	A. Human Vascular Tissue	7
	1. Analytical Procedure	7
	2. Results	7
	References	12
IV.	Carnosine	13
	A. Human Vascular Tissue	14
	1. Analytical Procedure	14
	2. Results	14
	References	20
V.	Coenzyme A	21
	A. Human Vascular Tissue	21
	1. Analytical Procedure	21
	a) Release of Pantothenic Acid	22
	b) Microbiological Determination of Pantothenic Acid	22
	2. Results	23
	B. Animal Vascular Tissue	26
	References	28
VI.	Creatine (Total Creatine)	30
	A. Human Vascular Tissue	30
	1. Analytical Procedure	30
	2. Results	34
	B. Animal Vascular Tissue	37
	References	39
VII.	Cytochrome c	40
	A. Human Vascular Tissue	40
	1. Analytical Procedure	40
	2. Results	41
	References	41

Contents

- VIII. Glutathione 42
  - A. Total Glutathione 42
    - 1. Human Vascular Tissue..................................... 42
      - a) Analytical Procedure 43
      - b) Results 43
    - 2. Animal Vascular Tissue.................................... 43
  - B. Reduced Glutathione 44
    - 1. Human Vascular Tissue..................................... 44
    - 2. Animal Vascular Tissue 44
  - References 48
- IX. Lipoic Acid 49
  - A. Human Vascular Tissue 49
    - 1. Analytical Procedure 49
      - a) Release of Protein-Bound Lipoic Acid 50
      - b) Microbiological Assay of Lipoic Acid 50
    - 2. Results 51
  - References 55
- X. Nucleotides and Nucleic Acids 56
  - A. Nucleotides 56
    - 1. Human Vascular Tissue 57
      - a) Analytical Procedure 57
      - b) Results 57
    - 2. Animal Vascular Tissue 58
  - B. Nucleic Acids 59
    - 1. Human Vascular Tissue.................................... 60
      - a) Analytical Procedure 60
      - b) Results 61
    - 2. Animal Vascular Tissue 61
  - References 72
- XI. Ubiquinone 76
  - A. Human Vascular Tissue 76
    - 1. Analytical Procedure 76
      - a) Saponification 76
      - b) Extraction of Ubiquinone 76
      - c) Combination of Ether Extracts 77
      - d) Washing of Combined Ether Extracts in Funnel No. 1 77
      - e) Evaporation of the Combined and Washed Ether Extract .. 77
      - f) Spectrophotometric Determination of Ubiquinone 77
      - g) Paper Chromatography of Ubiquinone 78
    - 2. Results 79
  - B. Animal Vascular Tissue 79
  - References 84

- Summary 85
- Subject Index 88

Acknowledgements

This research has been supported by the Washington University Ina Champ Urbauer Fund and by grants from Public Health Service (HE-00891), and the St. Louis Heart Association.

I am very grateful to Dr. J. R. CRISCIONE and Dr. J. J. THOMAS for their cooperation at the St. Louis City Morgue. My special thanks are due to my co-worker Dr. T. KHEIM, to my wife for the care and attention she has given to all details in the preparation of the manuscript, and to Mrs. M. BUCHANAN for her very accurate laboratory analyses.

Sincere thanks are also due to the following former members of our research group who worked with compounds included in this monograph: Dr. Y. O. CHANG, Dr. M. CHIEFFI, Dr. T. J. S. LAURSEN, Dr. R. SANWALD and Dr. I. WANG. I also want to thank the Muscular Dystrophy Association for supplying the special equipment for me to type the manuscript.

Finally, I wish to record my gratitude to the publisher, S. Karger AG, and to the coeditors of Atherosclerosis Monographs for valuable assistance.

List of Abbreviations

ADP	= adenosine diphosphate	GTP	= guanosine triphosphate
AMP	= adenosine monophosphate	IMP	= inosine monophosphate
ATP	= adenosine triphosphate	NAD	= nicotinamide-adenine dinucleotide
CDP	= cytidine diphosphate		
CMP	= cytidine monophosphate	NADP	= nicotinamide-adenine dinucleotidephosphate
CoA	= coenzyme A		
CTP	= cytidine triphosphate	PAPS	= 3'phosphoadenosine-5'-phosphosulfate
DNA	= deoxyribonucleic acid		
DNA-P	= deoxyribonucleic acid phosphorus	RNA	= ribonucleic acid
		RNA-P	= ribonucleic acid phosphorus
DTNB	= 5,5'-dithiobis-2-nitrobenzoic acid	UDP	= uridine diphosphate
		UDPgluc.	= uridine diphosphate glucuronic acid
GDP	= guanosine diphosphate		
GMP	= guanosine monophosphate	UMP	= uridine monophosphate
GSH	= reduced glutathione	UTP	= uridine triphosphate
GSSG	= oxidized glutathione		

Symbols

μg	= microgram	nm	= millimicron
ml	= milliliter	pg	= micromicrogram
mM	= millimolar	SD distr.	= standard deviation of distribution
MW	= molecular weight		
ng	= nanogram		

I. Introduction

There has been increasing agreement in recent years that investigation of arterial tissue metabolism may contribute to an understanding of the mechanism of atherogenesis. On the basis of quantitative biochemical studies performed during the last 25 years in the author's laboratory, the main metabolic pattern of human arteries has been established. This research has shown that the aortic wall has a low respiratory rate, a rather high rate of glycolysis and a low Pasteur effect [KIRK et al., 1954]. A comprehensive review of activities of 98 different enzymes in arterial tissue [KIRK, 1969] and a description of the general metabolism of the arterial wall have also been made [KIRK, 1968]. A systematic report of the concentrations in vascular tissue of compounds associated with tissue metabolism is also desirable.

An additional substance besides the enzyme and substrate is required in many cases in order that the reaction may proceed. The majority of enzyme cofactors may be divided broadly into two classes, specific coenzymes and activators. The activators are frequently of a very simple nature, e. g. inorganic ions which in various ways bring the enzyme itself into a catalytical state. In spite of their definite importance these inorganic cofactors are not included in this monograph which deals with specific coenzymes and some organic compounds that are of great significance for the metabolism of vascular tissue. Many coenzymes are derivatives of vitamins which have been described in a previous monograph [KIRK, 1973].

The word 'atherosclerosis' is interpreted by many scientists to mean only arterial lesions characterized by lipid deposits. Since separate pathological changes of connective tissue also often occur, and because it is considered important to conduct research separately on these two types of abnormal arterial tissue, the present author frequently uses the words 'lipid-arteriosclerotic' and 'fibrous-arteriosclerotic'.

This monograph contains quantitative values for several coenzymes and metabolic factors, including 3,803 assays of human vascular tissue and 1,865 determinations of animal vascular tissue; all available reliable data are reported. The majority of analyses of human vascular tissue have been performed by the author and his associates. The values are listed as (1) mean concentration for each decade of subjects with calculated SD distr.; the concentrations are expressed per gram wet tissue and per gram tissue nitrogen; (2) in addition, there are tables which compare contents of studied compounds in thoracic descending aortic tissue with those in the pulmonary artery, normal coronary artery, inferior vena cava and some other blood vessel samples from the same persons; because of the low susceptibility of the pulmonary artery and the vena cava to pathological changes, this information may be of some significance; (3) variations in compound concentrations with age have been determined systematically, most of the coefficients of correlation being calculated for the adult 20- to 89-year age group, and (4) tables are also provided in which the values for arteriosclerotic tissue portions are tabulated in percentage of the contents observed for normal segments of the same arterial samples. These latter tables make it possible to compare changes associated with aging and with the development of arteriosclerosis. In an initial chapter a survey is given of the effects of coenzymes and some metabolic factors on the development of atherosclerosis in animals receiving an atherogenic diet.

The coenzyme and metabolic factor measurements were made on vascular tissue from the St. Louis City Morgue obtained fresh at autopsy shortly after death. Only samples from persons without metabolic disease or infection were used. All assays were conducted on the intima-media layer of the blood vessels and were performed separately on normal and arteriosclerotic tissue portions; some analytical procedures are described in detail. For tissue homogenization a Kontes Dual grinder attached to a controlled electric stirrer was employed. Nitrogen determinations were made by the Kjeldahl method on sections of the same tissue specimens on which analyses were done.

References

KIRK, J. E.: Arteriosclerosis and arterial metabolism; in BITTAR and BITTAR The biological basis of medicine, vol. 1, pp. 493–519 (Academic Press, London 1968).
KIRK, J. E.: Enzymes of the arterial wall (Academic Press, New York 1969).
KIRK, J. E.: Vitamin contents of arterial tissue. Monogr. Atheroscler., vol. 3 (Karger, Basel 1973).
KIRK, J. E.; EFFERSØE, P. G., and CHIANG, S. P.: The rate of respiration and glycolysis by human and dog aortic tissue. J. Geront. *9:* 10–35 (1954).

II. Effects of some Coenzymes and Metabolic Factors on Experimental Atherosclerosis in Animals

Although the number of studies about the effects of coenzymes and special metabolic factors upon experimental atherosclerosis in animals produced by feeding them an atherogenic diet is somewhat limited, some of the findings are of great interest. In such investigations it is possible to determine the degree of atherosclerosis at autopsy. Control experiments have always been performed in which the animals received the same atherogenic diet without the special compound. A survey of published results is presented in table I.

The beneficial effect of cytochrome c administration to chickens fed an atherogenic diet reported in 1966 by YAMADA et al. [1966] may be in agreement with the earlier investigation by MARDONES et al. [1951] who studied rabbits receiving cholesterol-rich food and determined cytochrome c concentrations in the liver. It was found that animals who developed less atherosclerosis had higher cytochrome c content in the liver than those more susceptible to pathological aortic changes, the coefficient of correlation between liver cytochrome c level and developmental degree of atherosclerosis being -0.53 ($t = 2.88$).

Many factors may be responsible for the controversial results listed about the effectiveness of lipoic acid, namely differences in animal strain, duration of experiment, diet used and lipoic acid doses given.

It has been demonstrated by FILLIOS et al. [1958, 1959, 1960] that of the purines and pyrimidines studied, adenine and uracil displayed the highest enhancement of atherosclerosis; as seen from the table, only cytosine had no effect. An *in vitro* study by FLORENTIN et al. [1969] with labeled thymidine showed an increased incorporation of this nucleoside into endothelial cells of abdominal aorta from swine fed cholesterol for 3 days.

Although MILCH et al. [1952] found no certain effect of AMP on chickens with experimental atherosclerosis, a beneficial action of this nucleotide was observed when applied to old hens with spontaneous atherosclerosis.

In connection with the preventive action of ATP reported by GAMBASSI and MAGGI [1952, 1953 b, 1954 a, b] on rabbits fed an atherogenic diet it should be mentioned that HOSOKAWA and ARAKI [1964] subsequently have noted a therapeutic effect of ATP-cysteine when given to rabbits with arteriosclerosis induced by adrenaline and vitamin D. A survey of the ATP action on atherosclerosis has been published by GAMBASSI and MAGGI [1953 a].

An enhancement of experimental atherosclerosis in rats by administration of DNA was reported by FILLIOS *et al.* [1959] whereas, as listed in table I, no certain effect was found by THOMAS *et al.* [1963]. This difference may be due to the fact that in the study by THOMAS *et al.* only a single injection of 1.0 mg DNA was given.

Table I. A survey of reported effects of coenzymes and metabolic factors on atherosclerosis in animals placed on an atherogenic diet. The references in the table are listed for this chapter

Compound	Animal	Good preventive effect	No certain effect	Enhancement of atherosclerosis
Cytochrome c	rabbit	YAMADA *et al.*, 1966		
Lipoic acid	rabbit	ANGELUCCI and MASCITELLI-CORIANDOLI, 1958		KRITCHEVSKY and MOYER, 1958
Purines				
Adenine	rat			FILLIOS *et al.*, 1958, 1959
Guanine	rat			FILLIOS *et al.*, 1958, 1959
Xanthine	rat			FILLIOS *et al.*, 1958, 1959
Uric acid	rat			FILLIOS *et al.*, 1958, 1959
Pyrimidines				
Cytosine	rat		FILLIOS *et al.*, 1958, 1959	
Thymine	rat			FILLIOS *et al.*, 1958, 1959
Uracil	rat			FILLIOS *et al.*, 1958, 1959, 1960
Nucleotides				
AMP	chicken		MILCH *et al.*, 1952	
ATP	rabbit	GAMBASSI and MAGGI, 1952, 1953b, 1954a, b		
Nucleic acids				
DNA	rat		THOMAS *et al.*, 1963	FILLIOS *et al.*, 1959
RNA	rat			FILLIOS *et al.*, 1959

References

Angelucci, L. and Mascitelli-Coriandoli, E.: Anticholesterol activity of α-lipoic acid. Nature, Lond. *181:* 911–912 (1958).

Fillios, L. C.; Naito, C., and Andrus, S. B.: Purines and pyrimidines in experimental hypercholesteremia and atherogenesis. Fed. Proc. *17:* 436 (1958).

Fillios, L. C.; Naito, C.; Andrus, S. B., and Roach, A. M.: The hypercholesteremic and atherogenic properties of various purines and pyrimidines. Amer. J. clin. Nutr. *7:* 70–75 (1959).

Fillios, L. C.; Naito, C.; Andrus, S. B., and Roach, A. M.: Further studies on experimental atherosclerosis and dietary pyrimidines: Orotic acid, thiouracil, and uracil in male and female rats. Circulat. Res. *8:* 71–77 (1960).

Florentin, R. A. V.; Nam, S. C.; Lee, K. T., and Thomas, W. A.: Increased thymidine-^{3}H incorporation into endothelial cells of swine fed cholesterol for 3 days. Exp. molec. Path. *10:* 250–255 (1969).

Gambassi, G. e Maggi, V.: Aterosclerosi colesterolica e donatori di fosforo. Boll. Soc. ital. Biol. sper. *28:* 1493–1495 (1952).

Gambassi, G. e Maggi, V.: L'azione dell'ATP sul lipidi ematici ed in particolare sul rapporto fosfatidi-colesterolo, in normali e in dismetabolici di varia natura. Boll. Soc. ital. Biol. sper. *29:* 611–613 (1953a).

Gambassi, G. e Maggi, V.: L'ATP fattore di protezione dall'ateromasia sperimentale colesterolica: Il comportamento di alcune frazioni fosforate. Boll. Soc. ital. Biol. sper. *29:* 1650–1652 (1953b).

Gambassi, G. e Maggi, V.: Tempo di protrombina e di ricalcificazione nell'aterosclerosi sperimentale colesterolica: Contributo all'interpretazione del mecanismo antiateromasico dell'ATP. Acta geront. *4:* 73–80 (1954a).

Gambassi, G. e Maggi, V.: La lipasi tributirrinolitica e la malattia aterosclerotici sperimentale. Ricerca sul ruolo antiateromasico dell'ATP. Acta geront. *4:* 89–96 (1954b).

Hosokawa, S. and Araki, H.: Experimental studies on the treatment of arteriosclerosis. Treatment and prevention with adenosine triphosphate-cysteine. Yamaguchi Igaku *9:* 320–327 (1960); cit. Chem. Abstr. *60:* column 7339 (1964).

Kritchevsky, D. and Moyer, A. W.: Anticholesterol activity of α-lipoic acid. Nature, Lond. *182:* 396 (1958).

Mardones, J.; Monsalve, J.; Vial, M., and Plaza de los Reyes, M.: Tissue cytochrome c and prevention of experimental atherosclerosis. Science *114:* 387 (1951).

Milch, L. J.; Redmond, R. E.; Calhoun, W. W.; Chinn, H. I.; Bomar, O. W.; Johnson, C. W.; Yarnell, R. A.; Olson, R. L., and Paul, G.: Effects of adenosine-5-monophosphate on serum and aortal lipids in the chicken. Amer. J. Physiol. *170:* 346–350 (1952).

Thomas, W. A.; Jones, R.; Scott, R. F.; Morrison, E.; Goodale, F., and Imai, H.: Production of early atherosclerotic lesions in rats characterized by proliferation of 'modified smooth muscle cells'. Exp. molec. Path. .: suppl. 1, pp. 40–61 (1963).

Yamada, K.; Kuzuya, F., and Yamada, M.: Effects of cytochrome c on the experimental atherosclerosis of rabbits. Jap. Circulat. J. (Engl. Ed.) *30:* 895–898 (1966).

III. Carnitine (Free Carnitine)

Carnitine (β-hydroxy γ-butyrobetaine) is a compound which participates in important ways in lipid metabolism. Much information has been acquired about the functioning of carnitine after FRIEDMAN and FRAENKEL's discovery [1955] of carnitine acetyltransferase (EC acetyl-CoA: carnitine O-acetyltransferase; 2. 3. 1. 7) in animal tissue, an enzyme capable of reversibly acetylating carnitine from acetyl-CoA.

Carnitine stimulates the synthesis of lipids which predominantly occurs in the extramitochondrial space of cells and also enhances the rate of fatty acid oxidation in the mitochondria. Because CoA derivatives are not able to penetrate the mitochondrial membrane it is assumed that the metabolic action of carnitine to a great extent is due to its participation in mitochondrial translocation of acetyl groups. The acetyl group of intramitochondrial acetyl-CoA is transferred to carnitine to yield acetylcarnitine. This is followed by the diffusion of acetylcarnitine through the mitochondrial membrane and the transfer of the acetyl group from acetylcarnitine to extramitochondrial CoA. The resulting acetyl-CoA is then available for fatty acid synthesis. It has further been shown that the movement of long-chain fatty acids from extra- to intramitochondrial sites is accelerated by carnitine. In connection with this it should be mentioned that in a recent *in vitro* study by HASHIMOTO and DAYTON [1971] on normal rat aortic tissue it was found that the influence of carnitine on the oxidation of long-chain fatty acids was small as compared with other tissues; this is termed 'a unique behavior of aortic tissue' by these authors.

Similarities of its structure with acetylcholine and γ-aminobutyric acid have led to ideas about a role of carnitine in nerve and muscle function. Experiments made by HAYASHI [1965] on dogs showed that carnitine stimulated the motor pathway; it was further observed that when an excised skeletal muscle from a frog was placed in a Ringer solution to which carnitine had been added, the muscle produced rhythmical contractions for several minutes depending on the concentration of carnitine.

A preliminary report has been made by the author [KIRK, 1969] about free carnitine concentrations and carnitine acetyltransferase activities in human vascular tissue. A large number of such assays have been done subsequently and the free carnitine values acquired will be presented in this chapter.

A. Human Vascular Tissue

1. Analytical Procedure

For quantitative determination of free carnitine concentrations in vascular samples the method of MARQUIS and FRITZ [1964] was used, including their detailed description of the handling of tissues. In the author's opinion, before the final supernatant is obtained for enzymatic assay of carnitine, it is necessary to make careful measurements of tissue weight and of the volumes of the acquired solutions during the various steps involved in treating the tissue.

In the MARQUIS and FRITZ procedure acetyl-coenzyme A, DTNB and purified carnitine acetyltransferase are added to the sample; under the employed concentrations of these compounds carnitine receives acetyl from acetyl-CoA and the equimolar amount of liberated free CoA is measured spectrophotometrically by its combination with DTNB. A tissue blank and a reagent blank (containing only Tris-HCl buffer, DTNB and carnitine acetyltransferase) were run with each analysis.

For construction of the standard curve, L-carnitine (Mann Co., New York) was used; acetyl-CoA was obtained from Nutritional Biochemicals Corp. (Cleveland, Ohio) and purified carnitine acetyltransferase from Boehringer-Mannheim Co. (New York).

2. Results

The average carnitine values listed in table II show a mean concentration of 16.49 µg/g wet tissue in thoracic descending aorta. This content is about 6.2% of that reported by MARQUIS and FRITZ [1964] for rat cardiac tissue and 11.3% of rat skeletal muscle. According to ABDEL-KADER and WOLF [1965] the average carnitine content for rat serum is 5.1 µg/ml; the mean result for human aortic tissue is thus 3 times higher than that of rat serum.

A comparison of carnitine levels in various types of vascular samples from the same subjects (table III) displayed significantly higher concentrations in the pulmonary artery and coronary artery than in the thoracic descending aorta, whereas markedly lower values were found in the vena cava inferior and the iliac vein. The statistical calculations presented in table IV indicate a slight tendency of vascular tissue contents to decrease with age. Moderately lower carnitine values were recorded for fibrous-arteriosclerotic than for normal aortic tissue (table V).

Table II. Mean carnitine concentrations of human vascular tissue. Values expressed as micrograms of free carnitine per gram wet tissue and per gram tissue nitrogen

Vascular sample	Age group, years	Number	Wet tissue mean	SD distr.	Tissue nitrogen mean	SD distr.	Reference
Aorta normal[1]	0– 9	4	16.44		434.8		KIRK, 1969
	10–19	1	21.75		489.2		
	20–29	9	19.45	5.83	432.3	122.1	
	30–39	8	16.08	6.42	423.3	162.5	
	40–49	12	16.25	5.19	415.7	121.4	
	50–59	16	16.04	6.47	445.6	166.8	
	60–69	2	14.69		399.5		
	70–83	7	14.39	4.84	393.7	146.2	
	0–83	59	16.49	5.75	426.7	138.3	
	20–83	54	16.40	5.67	425.0	136.4	
Aorta, lipid-arteriosclerotic[1]	20–29	5	16.61	4.28	391.0	104.5	
	30–39	5	15.71	4.39	418.8	122.7	
	40–49	8	15.00	4.64	408.4	106.6	
	50–59	11	15.50	6.51	469.9	196.1	
	60–69	4	14.60		454.0		
	70–83	8	12.85	4.97	400.2	141.1	
	20–83	41	14.96	4.76	426.9	132.5	
Aorta, fibrous-arteriosclerotic[1]	40–49	6	15.15	8.47	405.5	205.8	
	50–59	9	13.26	3.91	386.5	109.5	
	60–69	4	14.76		407.3		
	70–83	5	13.99	6.02	410.4	155.4	
	40–83	24	14.13	5.27	399.7	133.3	

1 Thoracic descending aorta.

Table II (continued)

Vascular sample	Age group, years	Number	Wet tissue mean	SD distr.	Tissue nitrogen mean	SD distr.	Reference
Ascending aorta, normal	20–29	7	21.78	7.11	518.4	176.4	
	30–39	4	16.71		429.2		
	40–49	2	18.14		459.0		
	50–69	2	17.44		427.2		
	70–83	3	14.32		385.3		
	20–83	18	18.52	5.83	459.6	142.7	
Ascending aorta, arteriosclerotic	20–69	7	18.57	9.20	460.4	251.7	
Abdominal aorta, normal	19–29	9	18.87	4.82	530.1	112.0	
	30–39	3	13.81		382.7		
	50–59	1	15.86		397.6		
	19–59	13	17.47	4.76	485.9	110.5	
Abdominal aorta, arteriosclerotic	19–29	2	15.29		404.3		
	30–39	3	17.97		530.7		
	40–49	4	12.99		391.0		
	50–83	4	13.44		461.5		
	19–83	13	14.63	5.56	447.0	192.8	
Pulmonary artery, normal	0–9	3	12.04		338.7		
	10–19	1	10.41		264.8		
	20–29	8	20.73	10.46	518.9	245.1	
	30–39	7	17.52	11.36	474.6	268.3	
	40–49	9	18.00	7.43	503.0	180.8	
	50–59	10	20.91	10.16	586.3	252.9	
	60–69	6	15.89	4.22	473.0	105.8	
	70–83	8	14.88	4.52	452.4	144.6	
	0–83	52	17.70	8.16	492.3	201.0	
	20–83	48	18.21	8.29	506.7	202.2	
Coronary artery, normal	18–29	9	30.16	15.63	823.4	444.1	
	30–39	3	24.15		804.7		
	40–49	4	25.25		849.8		
	50–59	7	20.87	8.86	669.0	303.4	
	60–69	1	20.38		662.4		
	70–81	2	14.30		513.8		
	18–81	26	24.61	13.58	753.7	402.1	

Table II (continued)

Vascular sample	Age group, years	Number	Wet tissue mean	SD distr.	Tissue nitrogen mean	SD distr.	Reference
Coronary artery, arteriosclerotic	20–29	2	25.12		800.5		
	30–39	3	19.34		804.7		
	40–49	7	22.87	13.49	763.4	475.3	
	50–59	6	16.85	8.32	580.7	254.9	
	60–69	1	20.28		824.6		
	70–83	5	11.88	1.33	399.2	58.3	
	20–83	24	18.71	10.44	652.6	346.4	
Iliac artery, normal	19–29	8	15.21	5.18	428.3	146.7	
	30–39	5	17.01	3.08	482.6	80.7	
	40–59	4	16.62		500.3		
	19–59	17	16.07	6.43	461.2	178.0	
Iliac artery, arteriosclerotic	40–83	7	14.07	6.51	418.5	192.6	
Vena cava inferior	0–9	5	9.30	2.50	239.8	55.7	
	10–19	4	13.45		334.5		
	20–29	8	10.38	4.91	267.1	136.2	
	30–39	7	8.16	2.94	221.3	86.6	
	40–49	6	7.62	2.19	204.9	52.5	
	50–59	7	10.40	4.56	294.5	111.3	
	60–69	2	7.81		223.3		
	70–83	7	7.35	2.48	206.6	69.5	
	0–83	46	9.26	3.60	248.0	96.6	
	20–83	37	8.80	3.54	239.7	98.8	
Iliac vein	19–59	8	5.08	1.93	118.5	47.6	

Table III. Mean free carnitine concentrations of various types of normal vascular samples expressed in percent of contents of normal thoracic descending aortic tissue from the same subjects

Vascular sample	Age group, years	Number	Wet tissue %	t of difference	Tissue nitrogen %	t of difference	Reference
Ascending aorta, normal	20–83	16	102.9	0.49	106.2	0.97	KIRK. 1969
Abdominal aorta, normal	19–59	12	90.0	1.10	109.5	1.18	
Pulmonary artery, normal	0–83	41	115.4	2.22	121.1	2.73	
Coronary artery, normal	20–49	12	208.5	4.52	256.1	5.06	
	50–81	9	132.8	2.33	159.6	4.41	
	20–81	21	177.8	4.48	213.7	5.80	
Iliac artery, normal	19–59	15	94.4	0.48	115.5	1.06	
Vena cava inferior	0–83	38	54.5	8.66	58.1	7.22	
Iliac vein	19–59	8	24.1	10.55	24.1	10.76	

Table IV. Coefficients of correlation between age and tissue-free carnitine concentrations

Vascular sample	Age group, years	Number	Wet tissue r	t	Tissue nitrogen r	t	Reference
Aorta normal[1]	20–83	54	−0.25	1.88	−0.06	0.44	KIRK, 1969
Aorta, lipid-arteriosclerotic[1]	20–83	41	−0.24	1.55	+0.02	0.13	
Aorta, fibrous-arteriosclerotic[1]	40–83	24	−0.02	0.09	+0.04	0.21	
Pulmonary artery, normal	20–83	48	−0.17	1.17	−0.08	0.54	
Coronary artery, normal	18–81	26	−0.35	1.83	−0.23	1.16	
Coronary artery, arteriosclerotic	20–83	24	−0.37	1.86	−0.33	1.65	
Vena cava inferior	20–83	37	−0.20	1.21	−0.14	0.84	

1 Thoracic descending aorta.

Table V. Mean free carnitine concentrations of arteriosclerotic tissue expressed in percent of contents of normal tissue portions from the same arterial samples

Vascular sample	Age group, years	Number	Wet tissue %	t of difference	Tissue nitrogen %	t of difference	Reference
Aorta, lipid-arteriosclerotic[1]	20–49	18	91.4	1.80	96.3	0.54	KIRK, 1969
	50–83	20	96.8	0.66	106.6	0.89	
	20–83	38	94.1	1.58	101.8	0.40	
Aorta, fibrous-arteriosclerotic[1]	40–49	6	75.8	2.76	78.4	2.47	
	50–69	11	84.5	1.82	87.9	1.44	
	70–83	5	105.3	0.58	109.0	0.87	
	40–83	22	85.5	2.54	88.9	1.85	
Coronary artery, arteriosclerotic	20–81	11	96.6	0.50	102.2	0.36	

1 Thoracic descending aorta.

References

ABDEL-KADER, M. M. and WOLF, G.: The distribution of carnitine and its possible function in corticosteroid biosynthesis; in WOLF Recent research on carnitine, pp. 147–156 (MIT Press, Cambridge 1965).
FRIEDMAN, S. and FRAENKEL, G.: Reversible enzymatic acetylation of carnitine. Arch. Biochem. *59:* 491–501 (1955).
HASHIMOTO, S. and DAYTON, S.: Utilization of medium- and long-chain fatty acids by normal rat aorta, and the effect of DL-carnitine on their utilization. Atherosclerosis *13:* 345–354 (1971).
HAYASHI, T.: Physiological action of carnitine on nerve cells, nervous pathway, and muscle; in WOLF Recent research on carnitine, pp. 183–191 (MIT Press, Cambridge 1965).
KIRK, J. E.: Free carnitine content and carnitine acetyltransferase activity of human vascular tissue. J. Lab. clin. Med. *74:* 892 (1969).
MARQUIS, N. R. and FRITZ, I. B.: Enzymological determination of free carnitine concentrations in rat tissues. J. Lipid Res. *5:* 184–187 (1964).

IV. Carnosine

Carnosine is a dipeptide, β-alanyl-histidine, which as pointed out by SEVERIN *et al.* [1962] has attracted much attention in the field of biology. Although most textbooks of biochemistry and biology emphasize that the metabolic function of carnosine is not known, it should be mentioned that studies have been reported about the action of this dipeptide on the processes of glycolytic and oxidative phosphorylation [SEVERIN *et al.*, 1948] and it has also been shown in SEVERIN's laboratory that carnosine has a favorable effect on maintaining levels of high-energy-phosphorus compounds. Further investigations indicate that carnosine stabilizes the activity of pyruvate dehydrogenase [EC pyruvate: ferricytochrome b_1 oxidoreductase; 1. 2. 2. 2) and [NAGRADOVA, 1959] 3-phosphoglyceraldehyde dehydrogenase (EC D-glyceraldehyde-3-phosphate: NAD oxidoreductase [phosphorylating]; 1. 2. 1. 12). While BOCHARNIKOVA and PETUSHKOVA [1968] observed an inhibitory effect of carnosine on myosin ATPase (EC ATP phosphohydrolase; 3. 6. 1. 3) from animal muscular tissue AVENA and BOWEN [1969] found that the dipeptide stimulated the activity of this enzyme.

Several studies of carnosine concentration in muscular tissue have been published; a comprehensive review of values observed for 26 vertebrate species has recently appeared [CRUSH, 1970]. Information about the carnosine content in vascular tissue is not available in the literature. Although the actual metabolic function of carnosine has not yet been clearly established the present author decided to make extensive analyses of carnosine concentrations in human arterial and venous tissue with the belief that such quantitative data will be of importance when definite knowledge has been acquired about the metabolic action of this compound. A total of nearly 800 assays was made [KIRK, unpublished information].

A. Human Vascular Tissue

1. Analytical Procedure

For determination of carnosine in vascular samples the spectrophotometric method of PARKER [1966] was used including his description of preparation of tissue extracts. In this procedure carnosine reacts with diazotized *p*-bromoaniline to give a red color. The diazotization of *p*-bromoaniline was performed as outlined by KOESSLER and HANKE [1919]. Since, according to these authors, a prepared diazotized compound usually lasts only 24 h it was prepared fresh daily and the reagent was not used until 15 min or longer after the final dilution with water. A reagent blank was run with each analysis. For construction of a standard curve pure carnosine (Calbiochem Co., Los Angeles) was employed.

2. Results

The mean carnosine concentration found for normal thoracic descending aorta (table VI) was 0.6588 mg/g wet tissue; this is 23.8% of the average level reported by SEVERIN *et al.* [1962] for the proximal part of frog gastrocnemius muscle.

The carnosine values for human vascular samples in table VI reveal some interesting findings. One of these is the rather small variations in carnosine contents in each type of blood vessels studied (very low SD distr. for separately listed age groups). Another noteworthy observation is the higher carnosine concentrations in the tissues of the abdominal aorta, coronary artery and iliac artery than in the thoracic descending aorta; considerably higher values were recorded for the inferior vena cava and iliac vein. In the statistical calculations of these comparisons presented in table VII several of the t of differences are unusually high; the reasons for this is the low SD distr. for carnosine concentrations in the individual types of blood vessels.

A statistically significant increase in carnosine content with age for adult persons was demonstrated for arteriosclerotic ascending aortic tissue (table VIII) whereas a decrease was noted for the vena cava inferior. As seen from table IX, in elderly persons there was a slightly higher carnosine level in lipid-arteriosclerotic than in normal aortic tissue.

Table VI. Mean carnosine contents of human vascular tissue. Values expressed as milligrams of carnosine per gram wet tissue and per gram tissue nitrogen

Vascular sample	Age group, years	Number	Wet tissue mean	SD distr.	Tissue nitrogen mean	SD distr.	Reference
Aorta normal[1]	0– 9	5	0.7396	0.1225	16.48	1.58	KIRK, unpublished information
	10–19	6	0.6322	0.1732	15.83	4.38	
	20–29	9	0.6804	0.0829	17.25	1.91	
	30–39	15	0.5948	0.1400	16.30	3.62	
	40–49	18	0.6709	0.1513	17.55	3.85	
	50–59	27	0.6465	0.0797	17.58	1.90	
	60–69	9	0.6871	0.2234	19.84	6.43	
	70–82	5	0.7342	0.1198	19.60	2.79	
	0–82	94	0.6588	0.1236	17.49	3.20	
	20–82	83	0.6558	0.1204	17.67	3.24	
Aorta, lipid-arteriosclerotic[1]	18–29	8	0.6520	0.0842	17.23	2.26	
	30–39	5	0.6024	0.1002	17.74	3.96	
	40–49	9	0.6569	0.1055	18.26	3.38	
	50–59	22	0.7033	0.0954	20.84	2.67	
	60–69	7	0.6210	0.1157	17.72	4.84	
	70–82	8	0.7636	0.1000	22.65	3.45	
	18–82	59	0.6791	0.0975	19.57	3.09	
Aorta, fibrous-arteriosclerotic[1]	38–49	8	0.6675	0.1196	18.21	3.84	
	50–59	16	0.6596	0.1459	18.92	4.21	
	60–69	5	0.7196	0.0866	20.92	1.03	
	70–82	8	0.7163	0.1252	20.19	2.81	
	38–82	37	0.6816	0.1090	19.31	3.09	
Ascending aorta, normal	0– 9	3	0.5202		11.60		
	10–19	3	0.6836		16.37		
	20–29	5	0.7714	0.0866	20.19	1.12	
	30–39	10	0.6721	0.1103	17.25	2.49	
	40–49	12	0.6223	0.1241	16.51	2.68	
	50–59	15	0.7015	0.1100	18.52	2.41	
	60–69	5	0.6816	0.0864	18.34	1.66	
	70–82	6	0.6978	0.0894	18.12	1.89	
	0–82	59	0.6741	0.1064	17.52	2.57	
	20–82	53	0.6823	0.0951	17.92	2.23	
Ascending aorta, arteriosclerotic	30–39	4	0.5975		16.03		
	40–49	4	0.6720		17.72		

1 Thoracic descending aorta.

Table VI (continued)

Vascular sample	Age group, years	Number	Wet tissue mean	SD distr.	Tissue nitrogen mean	SD distr.	Reference
	50–59	14	0.7304	0.1174	19.14	1.92	
	60–69	6	0.6532	0.1072	17.45	2.90	
	70–82	6	0.7696	0.1264	22.06	3.82	
	30–82	34	0.7012	0.1044	18.82	2.63	
Abdominal aorta, normal	0– 9	6	0.8287	0.1856	20.30	2.57	
	10–19	8	0.8236	0.2033	23.66	5.32	
	20–29	9	0.8226	0.1000	24.37	2.32	
	30–39	9	0.7852	0.0527	23.51	2.51	
	40–49	9	0.6520	0.1131	18.81	3.10	
	50–59	5	0.7886	0.1455	23.44	4.12	
	70–81	2	0.7835		25.65		
	0–81	48	0.7794	0.1342	22.50	3.85	
	20–81	34	0.7602	0.1058	22.61	3.62	
Abdominal aorta, arteriosclerotic	18–29	6	0.8122	0.1118	24.13	3.06	
	30–39	8	0.7461	0.1134	23.19	3.78	
	40–49	12	0.7093	0.1225	21.21	3.20	
	50–59	15	0.7785	0.1513	25.03	4.66	
	60–69	5	0.6996	0.1249	22.74	2.74	
	70–81	5	0.8148	0.0902	25.22	2.61	
	18–81	51	0.7569	0.1217	23.53	3.82	
Pulmonary artery, normal	0– 9	2	0.6415		17.05		
	10–19	4	0.7557		22.05		
	20–29	5	0.6640	0.0935	19.48	2.96	
	30–39	12	0.6703	0.1264	19.13	3.62	
	40–49	15	0.6687	0.1000	19.26	2.37	
	50–59	26	0.5983	0.1316	17.60	2.85	
	60–69	9	0.5627	0.1118	17.04	2.84	
	70–82	9	0.6712	0.1274	20.10	3.96	
	0–82	82	0.6386	0.0949	18.66	2.71	
	20–82	76	0.6323	0.0946	18.52	2.69	
Coronary artery, normal	0– 9	3	1.1152		33.23		
	10–19	4	1.1779		35.73		
	20–29	6	0.9657	0.1342	29.57	2.76	
	30–39	10	0.8400	0.1749	26.66	5.94	
	40–49	10	0.7783	0.2236	23.84	6.42	
	50–59	12	0.8615	0.1568	26.96	4.36	

Table VI (continued)

Vascular sample	Age group, years	Number	Wet tissue mean	SD distr.	Tissue nitrogen mean	SD distr.	Reference
	60–69	3	0.9083		32.16		
	70–75	2	0.9700		30.30		
	0–75	50	0.9007	0.1805	28.11	5.45	
	20–75	43	0.8600	0.1676	27.05	5.14	
Coronary artery, arteriosclerotic	20–29	3	0.7980		26.37		
	30–39	7	0.9044	0.1353	30.24	5.85	
	40–49	7	0.8699	0.1472	28.00	5.82	
	50–59	20	0.7903	0.1170	26.76	3.61	
	60–69	8	0.7299	0.1252	24.65	4.58	
	70–82	8	0.8186	0.1307	29.41	3.80	
	20–82	53	0.8115	0.1049	27.44	4.26	
Iliac artery, normal	0–9	4	1.1403		27.83		
	10–19	5	0.7592	0.1658	21.62	3.67	
	20–29	6	0.8245	0.1843	23.90	4.17	
	30–39	10	0.7528	0.1293	23.16	3.78	
	40–49	11	0.7360	0.1205	22.83	3.35	
	50–60	9	0.8024	0.2344	24.98	4.86	
	0–60	45	0.8033	0.1752	23.78	3.86	
	20–60	36	0.7720	0.1574	23.64	3.75	
Iliac artery, arteriosclerotic	26–39	5	0.7088	0.1676	23.44	4.44	
	40–49	3	0.6807		21.63		
	50–59	12	0.8201	0.1206	26.75	3.23	
	60–69	8	0.6764	0.0781	23.26	3.17	
	70–82	5	0.8084	0.0902	25.20	2.24	
	26–82	33	0.7539	0.0919	24.70	3.06	
Vena cava inferior	0–9	6	1.5848	0.1844	37.98	4.51	
	10–19	8	1.2818	0.2948	35.61	6.93	
	20–29	9	1.2583	0.2121	33.31	5.03	
	30–39	13	1.1558	0.2235	32.42	6.97	
	40–49	15	1.1523	0.2737	33.25	7.20	
	50–59	24	1.0675	0.1562	30.33	4.58	
	60–69	8	1.0041	0.1461	28.76	3.78	
	70–82	9	1.0489	0.1414	30.90	4.42	
	0–82	92	1.1575	0.2385	32.27	5.74	
	20–82	78	1.1119	0.2151	31.49	5.30	

Table VI (continued)

Vascular sample	Age group, years	Number	Wet tissue mean	SD distr.	Tissue nitrogen mean	SD distr.	Reference
Iliac vein	0– 9	3	1.5290		35.17		
	10–19	3	1.5558		39.30		
	20–29	2	1.3121		33.95		
	30–39	5	1.1844	0.1061	32.56	3.04	
	40–49	7	1.2471	0.2796	35.64	7.73	
	50–59	14	1.2059	0.1412	34.04	3.60	
	60–69	5	1.1380	0.1765	31.20	3.63	
	70–79	4	1.1382		32.23		
	0–79	43	1.2478	0.2387	34.07	5.24	
	20–79	37	1.2000	0.2034	33.56	4.83	

Table VII. Mean carnosine concentrations of various types of normal vascular samples expressed in percent of contents of normal thoracic descending aortic tissue from the same subjects

Vascular sample	Age group, years	Number	Wet tissue %	t of difference	Tissue nitrogen %	t of difference	Reference
Ascending aorta, normal	0–82	57	105.4	1.36	101.4	0.52	KIRK, unpublished information
Abdominal aorta, normal	0–39	29	120.7	6.83	137.6	7.20	
	40–81	16	115.4	2.75	128.8	3.96	
	0–81	45	118.9	7.82	134.5	8.10	
Pulmonary artery, normal	0–39	23	109.8	3.02	120.5	4.28	
	40–49	15	97.6	0.05	106.8	1.57	
	50–82	40	91.7	2.21	98.4	0.57	
	0–82	78	97.9	0.44	106.0	1.69	
Coronary artery, normal	0–75	48	137.7	11.62	163.1	13.36	
Iliac artery, normal	0–60	42	124.6	5.85	143.1	9.56	
Vena cava inferior	0–19	11	211.0	9.66	234.2	11.21	
	20–49	36	181.9	12.59	191.3	12.25	
	50–59	24	165.7	13.96	172.5	14.90	

Table VII (continued)

	60–81	13	141.5	5.53	146.6	4.56
	0–81	84	174.6	17.28	183.2	18.22
Iliac vein	0–49	19	204.3	12.04	214.7	11.82
	50–75	20	169.3	11.40	185.0	11.33
	0–75	39	186.1	15.86	198.8	15.94

Table VIII. Coefficients of correlation between age and tissue carnosine concentrations

Vascular sample	Age group, years	Number	Wet tissue		Tissue nitrogen		Reference
			r	t	r	t	
Aorta normal[1]	20–82	83	—0.02	0.18	+0.15	1.37	KIRK, unpublished information
Aorta, lipid-arteriosclerotic[1]	18–82	59	+0.22	1.70	+0.32	2.54	
Aorta, fibrous-arteriosclerotic[1]	38–82	37	+0.17	1.02	+0.08	0.48	
Ascending aorta, normal	20–82	53	—0.08	0.57	—0.03	0.22	
Ascending aorta, arteriosclerotic	30–82	34	+0.35	2.12	+0.41	2.61	
Abdominal aorta, normal	20–81	34	—0.09	0.51	—0.02	0.11	
Abdominal aorta, arteriosclerotic	18–81	51	+0.02	0.14	+0.08	0.56	
Pulmonary artery, normal	20–82	76	—0.09	0.78	—0.08	0.69	
Coronary artery, normal	20–75	43	—0.05	0.32	+0.03	0.19	
Coronary artery, arteriosclerotic	20–82	53	—0.24	1.76	—0.07	0.50	
Iliac artery, normal	20–60	36	—0.05	0.29	+0.06	0.35	
Iliac artery, arteriosclerotic	26–82	33	+0.10	0.56	+0.10	0.56	
Vena cava inferior	20–82	78	—0.36	3.36	—0.23	2.06	
Iliac vein	20–79	37	—0.12	0.72	—0.06	0.35	

1 Thoracic descending aorta.

Table IX. Mean carnosine concentrations of arteriosclerotic tissue expressed in percent of contents of normal tissue portions from the same arterial samples

Vascular sample	Age group, years	Number	Wet tissue %	t of difference	Tissue nitrogen %	t of difference	Reference
Aorta, lipid-arteriosclerotic[1]	18–49	21	100.0	0.00	104.2	1.50	KIRK, unpublished information
	50–82	32	106.9	2.71	114.8	3.88	
	18–82	53	104.5	2.07	110.7	4.03	
Aorta, fibrous-arteriosclerotic[1]	38–82	34	103.4	1.17	106.9	1.66	
Ascending aorta, arteriosclerotic	35–82	17	101.3	0.34	102.8	0.63	
Abdominal aorta, arteriosclerotic	18–81	19	102.7	1.31	101.5	0.68	
Coronary artery, arteriosclerotic	20–49	11	97.3	0.75	99.5	0.07	
	50–75	13	84.0	3.39	88.7	1.65	
	20–75	24	90.0	2.90	93.6	1.49	
Iliac artery, arteriosclerotic	26–60	9	105.5	1.27	105.3	1.42	

1 Thoracic descending aorta.

References

AVENA, R. M. and BOWEN, W. J.: Effects of carnosine and anserine on muscle adenosine triphosphatases. J. biol. Chem. *244:* 1600–1604 (1969).

BOCHARNIKOVA, I. M. and PETUSHKOVA, E. V.: Comparison of the influence of natural imidazole compounds on the ATPase activity of myosins isolated from the muscles of various animals. Biokhimiya (Engl. Ed.) *33:* 599–601 (1968).

CRUSH, K. G.: Carnosine and related substances in animal tissues. Comp. Biochem. Physiol. *34:* 3–30 (1970).

KOESSLER, K. K. and HANKE, M. T.: Studies on proteinogenous amines. II. A microchemical colorimetric method for estimating imidazole derivatives. J. biol. Chem. *39:* 497–519 (1919).

NAGRADOVA, N. K.: A study of the properties of 3-phosphoglyceraldehyde dehydrogenase and soluble α-glycerophosphate dehydrogenase. Biokhimiya (Engl. Ed.) *24:* 315–323 (1959).

PARKER, C. J.: Spectrophotometric determination of carnosine and anserine in muscle. Analyt. chem. *38:* 1359–1362 (1966).

SEVERIN, S. E.; IVANOV, V. I.; KARUZINA, N. P., and YUDELOWICH, R. Y.: Action of carnosine on the carbohydrate-phosphorus muscle metabolism (in Russian). Biokhimiya *13:* 158–168 (1948).

SEVERIN, S. E.; VUL'FSON, P. L., and TRANDAFILOVA, L. L.: Carnosine content of different parts of frog muscles. Dokl. Akad. Nauk SSSR (Engl. Ed.) *145:* 663–665 (1962).

V. Coenzyme A

CoA (this means the coenzyme of acetylation) contains in its structure the vitamin pantothenic acid and the nucleotide adenosine-3',5'-diphosphate. It has been shown that virtually all the pantothenic acid present in tissues occurs in the form of CoA [NOVELLI et al., 1949]. CoA and its derivative acetyl-CoA occupy a key position in intermediary metabolism, being involved in both carbohydrate metabolism and fatty acid synthesis and oxidation. This coenzyme is greatly associated with transacetylation processes.

There are numerous enzyme reactions in which coenzyme A participates. A list of such enzymes has been made available by JAENICKE and LYNEN [1960]. It is interesting to note that 50 of these enzymes have been identified in animal tissues. CoA is essential for the initiation of the tricarboxylic acid cycle; the activity of the citrate condensing enzyme (EC citrate-oxalo-acetate-lyase CoA acetylating; 4. 1. 3. 7) in human vascular tissue has been reported [KIRK, 1966, 1969].

A. Human Vascular Tissue

Results of measured CoA concentrations in human vascular tissue have been published by SANWALD and KIRK [1964, 1965].

1. Analytical Procedure

Because human arterial samples usually are not available immediately after death, the rather simple acetylation procedure cannot be applied to these specimens. The reasons for this is that an appreciable portion of the CoA present in tissues is split quickly after death by autolysis with release of pantothenic acid. Quantitative measurement of CoA in human vascular tissue must, therefore, be conducted on the basis of microbiological pantothenic acid analysis.

a) Release of Pantothenic Acid

The procedure for enzymic release of pantothenic acid from coenzyme A was a modification of the technique outlined by NOVELLI [1955]. Aqueous homogenates of vascular samples were prepared using a Kontes Dual type tissue grinder. Quantitative liberation of pantothenic acid from tissue coenzyme A was achieved by treating the homogenates with intestinal alkaline phosphatase and with peptidase obtained from an acetone powder of hog kidney extract. Since it is very important to reduce the blank values in pantothenic acid determinations, only resin-treated kidney acetone powder was utilized (Pentex Inc., Kankakee, Ill.). The activity of the alkaline phosphatase preparation was assayed by the procedure of SCHMIDT and THANNHAUSER [1943], using 42 phosphatase u/g fresh vascular tissue. The efficiency of the hog kidney peptidase enzyme was determined as described by NOVELLI [1955] by testing its ability to release pantothenic acid from commercial pure CoA in the presence of an excess of alkaline phosphatase; 35 mg of hog kidney powder/g arterial tissue was found appropriate.

The enzymic treatment of the vascular homogenates was accomplished by incubating the samples in 0.135 M Tris buffer, pH 8.2 at 37° C for 6 h in a shaking water bath. A phosphatase and hog kidney powder blank in which the homogenate was substituted with a similar volume of redistilled water was run with each set of samples. At the end of the incubation period the tissue samples and blanks were diluted 1:10 with redistilled water, boiled for 3 min to stop the enzyme reaction and then filtered through Whatman No. 2 filter paper. The filtrates were subsequently centrifuged for 5 min at 25,000 r.p.m.

b) Microbiological Determination of Pantothenic Acid

For pantothenic acid measurement in the supernatants of the centrifuged tissue and blank samples the *Lactobacillus plantarum* (ATTC 8014) was employed as described by SKEGGS and WRIGHT [1944] and by The Association of Official Agricultural Chemists [1960]. Only newly purchased test tubes, carefully cleansed with redistilled water, were used in that procedure. Agar culture medium, inoculum broth and pantothenate assay medium were obtained from Difco Labs., Inc., Detroit, Mich. A pantothenic acid standard curve based on pantothenic acid contents of 0.01, 0.02, 0.04, 0.06 and 0.08 µg was prepared with each set of assays. All vascular samples and blanks were analyzed in triplicate at two different concentrations, requiring a total of 6 inoculated sample tubes for each microbiological pantothenic acid determination. The growth of *L. plantarum* was measured turbidimetrically after 20-hour incubation of the tubes at 37° C; the optical density readings were taken at 640 nm with a Beckman DU spectrophotometer. For calculation of coenzyme A on the basis of measured pantothenic acid the now generally accepted fact that 2.43 µg of CoA (1 Lipmann unit) contains 0.7 µg of pantothenic acid was used.

2. Results

The CoA concentrations of human vascular tissues (table X) are slightly and moderately higher than that reported by KAPLAN and LIPMANN [1948] for human red blood cells (7.3–9.7 µg/g) and must be definitely higher than in whole blood because coenzyme A was absent in plasma. In contrast to this, the values of both arterial and venous samples are lower than those found by PÜTTER and DARDENNE [1958] for normal eye lenses from 16 to 21- and 55 to 62-year-old subjects, the mean contents being, respectively, 33.3 and 21.1 µg/g wet tissue.

The average determinations listed in table X indicate a tendency toward an increase in the CoA level from the ascending to the abdominal aorta; however, this difference was demonstrated only for the abdominal aorta when comparison was made of aortic sections from the same adult persons (table XI). As seen from the same table, in the groups above 20 years of age statistically higher CoA concentrations were displayed by the pulmonary artery, coronary artery and inferior vena cava than by the thoracic descending aorta.

In some of the blood vessels analyzed, higher CoA values were found for children than for adults (table X). A notable increase in CoA with age was observed for the vena cava inferior samples from 20 to 59-year-old individuals (table XII) whereas no great variations were recorded for the various arteries studied.

The assays of arteriosclerotic and normal portions from the same arterial samples (table XIII) showed no significant decrease in coenzyme A content of lipid-arteriosclerotic aortic tissue whereas a markedly lower level was found for fibrous-arteriosclerotic specimens. The 11- to 15-percent lower concentrations recorded for lipid-arteriosclerotic than for normal coronary artery tissue and especially the statistically very high t of differences are of great interest.

Table X. Mean CoA contents of human vascular tissue. Values expressed as micrograms of coenzyme A per gram wet tissue and per gram tissue nitrogen

Vascular Sample	Age group, years	Number	Wet tissue mean	SD distr.	Tissue nitrogen mean	SD distr.	Reference
Aorta normal[1]	0–9	10	16.25	7.56	348.6	162.7	SANWALD
	10–19	1	13.52		266.7		and KIRK,
	20–29	15	9.92	4.10	231.0	91.2	1964, 1965
	30–39	17	9.14	3.74	226.8	92.8	
	40–49	17	11.84	5.31	308.7	118.6	
	50–59	24	12.81	5.69	338.1	134.8	
	60–69	9	9.71	5.92	264.6	165.9	
	70–79	5	11.38	4.81	338.0	147.4	
	80–85	2	11.24		308.5		
	0–85	100	11.55	5.48	291.1	139.2	
	20–85	89	11.03	5.21	285.0	132.5	
Aorta, lipid-arteriosclerotic[1]	20–29	4	11.68		317.1		
	30–39	4	9.47		287.7		
	40–49	4	9.62		300.3		
	50–59	6	12.26	4.87	357.0	143.0	
	60–69	8	9.64	4.70	279.3	127.1	
	70–85	3	5.97		199.5		
	20–85	29	10.06	4.59	296.5	131.9	
Aorta, fibrous-arteriosclerotic[1]	40–49	2	7.62		205.8		
	50–59	5	6.15	4.20	191.1	128.1	
	60–69	1	2.86		92.4		
	70–85	1	3.84		111.3		
	40–85	9	5.85	4.06	174.1	114.9	
Ascending aorta, normal	0–9	1	11.97		241.5		
	20–29	2	8.78		212.1		
	30–39	2	7.85		197.4		
	40–49	1	10.52		279.3		
	50–59	2	8.86		226.8		
	0–59	8	9.19	1.70	223.5	40.9	
	20–59	7	8.79	1.39	220.7	43.6	
Abdominal aorta, normal	0–9	4	19.09		430.5		
	20–29	6	12.79	4.20	329.7	104.4	
	30–39	1	20.88		611.1		
	40–49	2	16.07		465.2		
	50–59	4	13.52		350.7		

1 Thoracic descending aorta.

Table X (continued)

Vascular sample	Age group, years	Number	Wet tissue mean	SD distr.	Tissue nitrogen mean	SD distr.	Reference
	0–59	17	15.31	5.71	390.8	140.9	
	20–59	13	14.14	4.83	378.6	128.5	
Abdominal aorta, lipid-arteriosclerotic	20–29	4	11.89		354.9		
	30–39	1	10.23		363.3		
	50–59	2	14.33		443.1		
	20–59	7	12.35	4.09	381.4	118.6	
Pulmonary artery, normal	0–9	2	17.83		485.1		
	10–19	1	16.99		445.2		
	20–29	7	13.76	5.69	369.6	142.3	
	30–39	5	14.43	4.07	373.8	102.7	
	40–49	4	16.76		510.3		
	50–59	9	15.81	6.24	447.3	170.1	
	60–69	4	11.66		363.4		
	70–79	1	8.80		262.5		
	0–79	33	14.76	4.94	413.7	132.1	
	20–79	30	14.43	5.08	407.8	138.0	
Coronary artery, normal	0–9	2	19.95		552.3		
	20–29	6	14.67	5.14	426.1	151.2	
	30–39	7	16.35	5.25	436.8	138.8	
	40–49	1	24.96		869.4		
	50–59	8	17.95	6.48	495.6	176.4	
	60–69	1	17.54		494.8		
	0–69	25	17.14	5.15	481.9	152.9	
	20–69	23	16.89	5.21	475.8	154.3	
Coronary artery, lipid-arteriosclerotic	20–29	3	16.07		496.2		
	30–39	2	13.42		420.0		
	40–49	2	19.60		542.8		
	50–59	5	11.88	4.45	409.1	148.5	
	70–85	2	20.67		551.3		
	20–85	14	15.36	4.52	468.7	143.9	
Vena cava inferior	0–9	2	22.58		525.1		
	20–29	5	12.31	0.58	310.8	35.9	
	30–39	2	18.19		498.7		
	40–49	2	12.09		317.1		
	50–59	3	18.90		513.0		
	0–59	14	16.00	4.99	412.2	115.5	
	20–59	12	14.90	4.60	393.4	116.5	

Table XI. Mean CoA concentrations of normal ascending aorta, normal abdominal aorta, pulmonary artery, normal coronary artery, and vena cava inferior samples expressed in percent of contents of normal thoracic descending aortic tissue from the same subjects

Vascular sample	Age group, years	Number	Wet tissue %	t of difference	Tissue nitrogen %	t of difference	Reference
Ascending aorta, normal	0–59	8	104.6	0.43	102.5	0.21	SANWALD and KIRK, 1964, 1965
Abdominal aorta, normal	0– 9	4	80.5	1.08	82.9	1.26	
	20–59	13	116.7	1.80	127.9	3.34	
	0–59	17	103.1	0.35	112.1	1.36	
Pulmonary artery, normal	0–19	3	74.4	1.27	90.8	0.46	
	20–49	16	136.6	3.06	145.7	3.90	
	50–79	13	144.8	4.62	153.9	6.55	
	0–79	32	127.4	3.24	139.4	5.47	
Coronary artery, normal	0– 9	2	69.3	1.50	88.1	0.50	
	20–39	13	163.1	3.00	184.3	3.58	
	40–69	10	207.6	5.93	219.9	5.67	
	0–69	25	157.9	4.54	180.0	5.09	
Vena cava inferior	0– 9	2	78.5	5.32	85.8	2.13	
	20–59	12	132.7	3.11	139.5	3.53	
	0–59	14	130.1	2.87	125.7	2.53	

B. Animal Vascular Tissue

Essential investigations of CoA in animal aortic tissues and its relationship to lipid synthesis and atherogenesis have been conducted at Milan University, Italy, by PAOLETTI *et al.* [1959] and by KIM *et al.* [1960]. The mean CoA values (No. 6 of each set of assays) for adult animals when expressed as micrograms per gram wet tissue were: pigeon, 22.56; rat, 23.57; hamster, 24.54; guinea pig, 29.40; rabbit, 42.53; and chicken, 43.98. The corresponding CoA contents for young rats and rabbits were both 18.47 µg/g wet tissue which indicates a distinctly lower level in young than in adult rabbits.

Table XII. Coefficients of correlation between age and tissue CoA concentrations

Vascular sample	Age group, years	Number	Wet tissue		Tissue nitrogen		Reference
			r	t	r	t	
Aorta normal[1]	0–85	100	—0.09	0.90	+0.07	0.69	SANWALD
	20–85	89	+0.13	1.22	+0.21	2.00	and KIRK, 1964, 1965
Aorta, lipid-arteriosclerotic[1]	20–85	29	—0.16	0.85	+0.04	0.21	
Aorta, fibrous-arteriosclerotic[1]	40–85	9	—0.25	0.68	—0.24	0.65	
Ascending aorta, normal	0–59	8	—0.38	1.01	—0.09	0.22	
	20–59	7	+0.17	0.39	+0.17	0.39	
Abdominal aorta, normal	0–59	17	—0.27	1.10	—0.10	0.39	
	20–59	13	+0.02	0.07	+0.05	0.17	
Pulmonary artery, normal	0–79	33	—0.11	0.57	0.00	0.00	
	20–79	30	—0.03	0.16	—0.01	0.05	
Coronary artery, normal	0–69	25	+0.19	0.92	+0.20	0.97	
	20–69	22	+0.20	0.93	+0.28	1.34	
Coronary artery, lipid-arterio-sclerotic	20–85	14	+0.18	0.63	+0.11	0.38	
Vena cava inferior	0–59	14	—0.12	0.42	+0.05	0.17	
	20–59	12	+0.48	1.73	+0.39	1.34	

1 Thoracic descending aorta.

A study of the CoA variation with age in swine aortic tissue (intima plus media plus adventitia) has been mady by ALBERTINI et al. [1967]; these assays displayed an average value for 5- to 6-year-old animals, which was 67.4% of that recorded for 1- to 16-month-old swines. A decrease in CoA was observed by MANTA et al. [1961] in rabbits fed an atherogenic diet.

A very important observation which was pointed out by PAOLETTI et al. [1959] is that animal species which are most susceptible to arteriosclerosis have the highest coenzyme A concentrations in aortic tissue. It was further demonstrated by *in vitro* experiments that animal aortas with the most elevated CoA levels also had the highest rate of lipid synthesis from ^{14}C-labeled acetate.

Table XIII. Mean CoA concentrations of arteriosclerotic tissue expressed in percent of contents of normal tissue portions from the same arterial samples

Vascular sample	Age group, years	Number	Wet tissue %	t of difference	Tissue nitrogen %	t of difference	Reference
Aorta, lipid-arteriosclerotic[1]	20–49	12	96.5	0.27	109.0	0.56	SANWALD and KIRK, 1964, 1965
	50–85	17	93.0	0.84	98.9	0.12	
	20–85	29	94.5	0.79	102.6	0.32	
Aorta, fibrous-arteriosclerotic[1]	20–85	9	54.5	4.50	64.5	3.50	
Abdominal aorta, arteriosclerotic	20–59	6	88.6	1.48	106.1	0.67	
Coronary artery, lipid-arterio-sclerotic	20–39	5	88.8	8.56	93.6	4.44	
	50–59	5	84.8	6.00	93.4	1.48	
	20–59	10	87.0	9.38	93.5	3.09	

1 Thoracic descending aorta.

In connection with these findings it is interesting to note that the CoA analyses performed by SANWALD and KIRK [1964, 1965] revealed much lower contents in human arterial tissues than those listed for animal aortas.

References

ALBERTINI, E.; MARTINESI, L. e BELLI, C.: Indagine sperimentale di elementi strutturali e biochimici arterologici nella specie 'Sus domestica' in funzione di stadi di età. G. Clin. med. *48:* 66–79 (1967).
Association of Official Agricultural Chemists: Official method of analysis, 9th ed. (1960).
JAENICKE, L. and LYNEN, F.: Coenzyme A; in BOYER, LARDY and MYRBÄCK The enzymes, vol. 3, pp. 3–103 (Academic Press, New York 1960).
KAPLAN, N. O. and LIPMANN, F.: The assay and distribution of coenzyme A. J. biol. Chem. *174:* 37–44 (1948).
KIM, J. J.; PAOLETTI, R., and VERTUA, R.: An autoradiographic approach to cholesterol biosynthesis and metabolism. Atompraxis *6* (2): 55–58 (1960).
KIRK, J. E.: Citrate condensing enzyme activity of human arterial and venous tissue. J. Lab. clin. Med. *68:* 888–889 (1966).
KIRK, J. E.: Citrate condensing enzyme; in KIRK Enzymes of the arterial wall, pp. 397–402 (Academic Press, New York 1969).

MANTA, I.; BEDELEANU, D.; CAPILNA, S.; MURESAN, L., and GORUN, V.: On certain biochemical changes in experimental atherosclerosis. Roum. med. Rev. *5:* 181–182 (1961).
NOVELLI, G. D.: Methods for determination of coenzyme A. Meth. biochem. Anal. *2:* 189–213 (1955).
NOVELLI, G. D.; KAPLAN, N. O., and LIPMANN, F.: The liberation of pantothenic acid from coenzyme A. J. biol. Chem. *177:* 97–107 (1949).
PAOLETTI, P.; TESSARI, L. e VERTUA, R.: Studio delle biosintesi lipidiche aortiche in speci animali diverse. Ric. sci. *29:* 2382–2389 (1959).
PÜTTER, J. und DARDENNE, U.: Der Gehalt an Coenzym A in jungen, alten und kataraktösen Augenlinsen. Z. Physiol. Chem. *310:* 59–64 (1958).
SANWALD, R. and KIRK, J. E.: Coenzyme A content of human arterial and venous tissue. Fed. Proc. *23* (2):145 (1964).
SANWALD, R. and KIRK, J. E.: Coenzyme A content of human arterial and venous tissue. J. Atheroscler. Res. *5:* 497–503 (1965).
SCHMIDT, G. and THANNHAUSER, S. J.: Intestinal phosphatase. J. biol. Chem. *149:* 369–389 (1943).
SKEGGS, H. R. and WRIGHT, L. D.: The use of *Lactobacillus arabinosus* in the microbiological determination of pantothenic acid. J. biol. Chem. *156:* 21–26 (1944).

VI. Creatine (Total Creatine)

Creatine (α-methylguanidine) is present in tissues both phosphorylated as creatine phosphate and in the free state. Creatine phosphate is an important constituent of living cells because it is an energy-rich phosphate derivative; it is an essential component of chemical reactions which produce muscular contraction since it assures the maintenance of a maximum supply of ATP. The transfer of high-energy phosphate from creatine phosphate to ADP, resulting in the formation of ATP is catalyzed by the creatine phosphokinase enzyme (EC ATP: creatine phosphotransferase; 2. 7. 3. 2); the reaction is reversible:

$$\text{creatine phosphate} + \text{ADP} \rightleftarrows \text{creatine} + \text{ATP}.$$

Rather high activities of this enzyme have been demonstrated in human vascular tissue [KIRK, 1962, 1963, 1964, 1969]. It may be assumed that the creatine phosphate compound is of great metabolic importance regarding the physiological functioning of vascular tissue.

A. Human Vascular Tissue

Measurements of total creatine in human vascular tissue have recently been made by the present author [KIRK, unpublished information]. Because creatine phosphate and free creatine are converted to each other in tissue samples it was considered advisable to determine total creatine contents.

1. Analytical Procedure

For assay of total creatine a tissue homogenate was incubated with pure creatine phosphokinase (Boehringer-Mannheim Co., New York), ADP and magnesium chloride in Tris buffer (pH 6.8), using the modification of CHAPPELL and PERRY's procedure [1954] as previously described in detail [KIRK, 1962, 1969]. Aliquots were removed before addition of creatine phosphokinase and after 10-min incubation at 37°C in a shaking water bath. Each aliquot was added to neutralized phenylmercuric acetate solution (in 50% dioxane) in a test tube. For protein precipitation barium hydroxide and zinc sulfate reagents were used.

After centrifugation the resulting supernatants were available for colorimetric assay of free creatine. The method of EGGLETON *et al.* [1943] was employed in a slightly modified from [KIRK, 1962, 1969] for spectrophotometric creatine determination. A reagent blank in which the tissue homogenate was replaced with an equal volume of redistilled water was run with each test. The efficiency of the procedure for complete release of free creatine was ascertained by conducting measurements on solutions of pure creatine phosphate (Boehringer-Mannheim Co., New York).

Table XIV. Mean creatine contents of human vascular tissue. Values expressed as micromoles of total creatine per gram wet tissue and per gram tissue nitrogen

Vascular sample	Age group, years	Number	Wet tissue mean	SD distr.	Tissue nitrogen mean	SD distr.	Reference
Aorta normal[1]	0– 9	11	1.270	0.468	31.03	12.19	KIRK, unpublished information
	10–19	7	1.310	0.295	30.87	5.20	
	20–29	18	1.080	0.396	27.05	8.75	
	30–39	17	1.055	0.443	27.73	10.55	
	40–49	32	0.802	0.318	22.07	8.51	
	50–59	30	0.752	0.298	20.89	7.10	
	60–69	28	0.649	0.163	18.35	4.30	
	70–86	11	0.565	0.252	16.44	6.65	
	0–86	154	0.865	0.370	23.01	9.58	
	20–86	136	0.809	0.351	21.95	9.26	
Aorta, lipid-arteriosclerotic[1]	20–29	5	0.645	0.151	16.25	4.58	
	30–39	5	0.540	0.158	14.75	3.97	
	40–49	7	0.412	0.219	11.37	5.94	
	50–59	13	0.392	0.122	11.49	3.42	
	60–69	9	0.396	0.203	12.40	6.02	
	70–86	7	0.416	0.158	13.14	5.83	
	20–86	46	0.443	0.183	12.77	4.97	
Aorta, fibrous-arteriosclerotic[1]	40–49	8	0.510	0.276	14.35	5.05	
	50–59	5	0.451	0.108	13.74	2.82	

1 Thoracic descending aorta.

Table XIV (continued)

Vascular sample	Age group, years	Number	Wet tissue mean	SD distr.	Tissue nitrogen maen	SD distr.	Reference
	60–69	7	0.418	0.167	12.12	4.15	
	70–84	4	0.383		10.16		
	40–84	24	0.450	0.182	12.87	4.94	
Ascending aorta, normal	0–9	3	0.738		18.19		
	10–19	3	0.947		23.11		
	20–29	9	0.853	0.264	20.26	5.06	
	30–39	6	0.726	0.167	17.70	3.46	
	40–49	9	0.603	0.160	15.70	3.95	
	50–59	4	0.775		20.58		
	60–69	3	0.557		15.74		
	70–80	4	0.568		15.48		
	0–80	41	0.721	0.231	18.17	5.28	
	20–80	35	0.700	0.228	17.75	5.14	
Ascending aorta, arteriosclerotic	20–29	1	0.456		10.52		
	30–39	3	0.650		16.32		
	40–49	3	0.593		13.80		
	50–59	4	0.627		18.13		
	60–69	2	0.407		10.65		
	70–80	3	0.518		14.69		
	20–80	16	0.567	0.188	14.92	4.82	
Abdominal aorta, normal	0–9	2	1.110		30.42		
	10–19	1	0.775		22.02		
	20–29	4	0.801		22.73		
	30–39	3	0.895		27.82		
	40–49	4	0.805		22.42		
	50–59	1	1.128		33.28		
	70–78	1	0.205		6.34		
	0–78	16	0.840	0.341	24.16	9.50	
	20–78	13	0.803	0.322	23.36	9.11	
Abdominal aorta, arteriosclerotic	20–29	4	0.594		17.23		
	30–39	4	0.769		24.22		
	40–49	10	0.556	0.232	17.18	6.01	
	50–59	3	0.542		20.17		
	60–69	2	0.588		19.16		
	70–80	4	0.288		9.10		
	20–80	27	0.554	0.266	17.51	8.10	

Table XIV (continued)

Vascular sample	Age group, years	Number	Wet tissue mean	SD distr.	Tissue nitrogen mean	SD distr.	Reference
Pulmonary artery, normal	0–9	2	0.876		23.91		
	10–19	1	1.326		38.75		
	20–29	5	0.917	0.269	24.56	6.33	
	30–39	6	1.005	0.168	28.02	4.82	
	40–49	14	0.995	0.326	29.02	9.88	
	50–59	14	1.078	0.287	31.64	7.02	
	60–69	7	0.812	0.212	23.42	6.06	
	70–84	7	0.744	0.238	21.80	6.70	
	0–84	56	0.957	0.243	27.56	7.42	
	20–84	53	0.953	0.245	27.49	7.45	
Coronary artery, normal	0–9	1	1.227		38.32		
	10–19	2	1.225		37.13		
	20–29	5	1.035	0.632	29.02	17.35	
	30–39	5	0.886	0.269	29.30	7.33	
	40–49	5	0.657	0.296	20.58	8.45	
	50–59	5	0.667	0.170	22.17	4.06	
	60–69	3	0.779		26.96		
	70–84	2	0.844		26.29		
	0–84	28	0.855	0.370	26.84	9.86	
	20–84	25	0.810	0.346	25.55	9.14	
Coronary artery, arteriosclerotic	20–29	1	0.822		28.53		
	30–39	4	0.580		19.23		
	40–49	7	0.512	0.196	19.21	8.18	
	50–59	3	0.424		17.90		
	60–69	5	0.309	0.217	11.03	7.14	
	70–80	2	0.322		11.07		
	20–80	22	0.463	0.265	16.86	8.44	
Iliac artery, normal	0–9	2	0.998		25.26		
	10–19	1	0.923		27.00		
	20–29	7	0.819	0.281	23.95	6.20	
	30–39	3	0.973		30.68		
	40–49	6	0.630	0.195	19.17	5.15	
	50–59	2	0.974		30.70		
	60–69	1	0.719		23.22		
	70–80	3	0.616		19.90		
	0–80	25	0.792	0.331	23.86	9.38	
	20–80	22	0.768	0.314	23.59	9.06	

Table XIV (continued)

Vascular sample	Age group, years	Number	Wet tissue mean	SD distr.	Tissue nitrogen mean	SD distr.	Reference
Iliac artery, arteriosclerotic	20–29	1	0.960		29.94		
	30–39	3	0.742		24.36		
	40–49	10	0.553	0.149	17.09	4.16	
	50–59	4	0.534		18.66		
	60–69	2	0.619		19.36		
	70–80	4	0.408		12.91		
	20–80	24	0.572	0.248	18.29	7.64	
Vena cava inferior	0–9	3	0.490		12.05		
	10–19	1	0.482		11.82		
	20–29	7	0.448	0.216	11.97	5.01	
	30–39	3	0.575		16.83		
	40–49	14	0.456	0.154	12.95	3.88	
	50–59	5	0.362	0.163	10.40	3.50	
	60–69	3	0.350		10.17		
	70–84	4	0.247		7.58		
	0–84	40	0.426	0.184	11.91	4.82	
	20–84	36	0.419	0.189	11.90	5.02	
Iliac vein	0–9	1	0.418		9.42		
	10–19	1	0.221		6.51		
	20–29	2	0.399		10.78		
	30–39	2	0.257		7.85		
	40–49	8	0.347	0.145	9.51	2.81	
	50–59	4	0.240		6.45		
	60–69	2	0.265		7.23		
	70–73	1	0.256		7.07		
	0–73	21	0.308	0.107	8.41	3.01	
	20–73	19	0.307	0.109	8.45	3.13	

2. Results

The mean creatine content listed in table XIV for normal thoracic descending aorta (0.865 µM/g wet tissue) is definitely higher than the average values for human whole blood reported by SUTTON *et al.* [1954] and by PILSUM *et al.* [1956] who found, respectively, 0.120 and 0.206 µM/ml blood.

Table XV. Mean creatine concentrations of various types of normal vascular samples expressed in percent of contents of normal thoracic aortic descending tissue from the same subjects

Vascular sample	Age group, years	Number	Wet tissue %	t of difference	Tissue nitrogen %	t of difference	Reference
Ascending aorta, normal	0–39	16	98.9	0.25	92.9	1.30	Kirk, unpublished information
	40–80	18	93.7	0.74	91.8	1.15	
	0–80	34	96.3	0.75	92.3	1.70	
Abdominal aorta, normal	0–59	14	106.9	0.47	118.5	1.58	
Pulmonary artery, normal	0–49	25	134.2	4.98	143.4	5.52	
	50–84	23	148.4	5.53	151.1	5.86	
	0–84	48	140.6	7.40	147.1	8.23	
Coronary artery, normal	0–39	11	110.5	0.55	136.0	2.34	
	40–84	15	108.7	1.21	124.3	2.64	
	0–84	26	109.5	1.20	129.3	3.51	
Iliac artery, normal	0–80	20	96.4	0.86	109.1	1.03	
Vena cava inferior	0–39	12	60.8	3.42	64.9	3.23	
	40–49	11	59.6	4.94	62.2	4.93	
	50–84	10	44.5	6.02	45.9	5.28	
	0–84	33	55.4	8.65	57.8	7.37	
Iliac vein	0–39	6	49.0	3.83	52.6	2.63	
	40–49	6	44.9	7.74	45.2	11.18	
	50–73	6	35.8	5.92	34.4	5.89	
	0–73	18	43.2	9.88	43.7	8.22	

The results of the many analyses and statistical calculations presented in tables XIV, XV, XVI and XVII reveal some important findings. As seen from table XV, there were notably higher creatine concentrations in pulmonary artery samples than in thoracic descending aortic tissue from the same subjects whereas markedly lower values were displayed by the vena cava inferior and the iliac vein.

Table XVI. Coefficients of correlation between age and tissue total creatine concentrations

Vascular sample	Age group, years	Number	Wet tissue		Tissue nitrogen		Reference
			r	t	r	t	
Aorta normal[1]	0–86	154	—0.58	8.76	—0.48	6.77	KIRK, unpublished information
	20–86	136	—0.49	6.58	—0.42	5.38	
Aorta lipid arteriosclerotic[1]	20–86	46	—0.37	2.67	—0.20	1.36	
Aorta, fibrous-arteriosclerotic[1]	40–84	24	—0.40	2.05	—0.33	1.65	
Ascending aorta, normal	0–80	41	—0.47	3.32	—0.34	2.26	
	20–80	35	—0.43	2.73	—0.32	1.95	
Ascending aorta, arteriosclerotic	20–80	16	—0.19	0.73	—0.04	0.15	
Abdominal aorta, normal	0–78	16	—0.41	1.70	—0.39	1.58	
	20–78	13	—0.41	1.49	—0.38	1.37	
Abdominal aorta, arteriosclerotic	20–80	27	—0.49	2.80	—0.34	1.82	
Pulmonary artery, normal	0–84	56	—0.18	1.35	—0.13	0.97	
	20–84	53	—0.16	1.16	—0.16	1.16	
Coronary artery, normal	0–84	28	—0.40	2.22	—0.35	1.92	
	20–84	25	—0.31	1.57	—0.20	1.01	
Coronary artery, arteriosclerotic	20–80	22	—0.52	2.73	—0.54	2.89	
Iliac artery, normal	0–80	25	—0.32	1.62	—0.16	0.78	
	20–80	22	—0.22	1.02	—0.13	0.60	
Iliac artery, arteriosclerotic	20–80	24	—0.43	2.24	—0.44	2.32	
Vena cava inferior	0–84	40	—0.34	2.24	—0.24	1.55	
	20–84	36	—0.36	2.27	—0.30	1.83	
Iliac vein	0–73	21	—0.23	1.04	—0.21	0.94	
	20–73	19	—0.28	1.21	—0.30	1.30	

1 Thoracic descending aorta.

Table XVII. Mean creatine concentrations of arteriosclerotic tissue expressed in percent of contents of normal tissue portions from the same arterial samples

Vascular sample	Age group, years	Number	Wet tissue %	Wet tissue t of difference	Tissue nitrogen %	Tissue nitrogen t of difference	Reference
Aorta, lipid-arteriosclerotic[1]	20–49	15	67.5	8.25	69.5	7.76	KIRK, unpublished information
	50–59	13	64.2	6.40	64.9	5.58	
	60–86	14	61.2	6.75	66.8	6.73	
	20–86	42	64.5	12.32	67.3	11.90	
Aorta, fibrous-arteriosclerotic[1]	40–84	22	65.9	7.62	67.0	8.34	
Ascending aorta, arteriosclerotic	20–80	13	69.9	4.21	74.0	3.90	
Abdominal aorta, arteriosclerotic	20–78	10	76.3	3.66	80.2	2.12	
Coronary artery, arteriosclerotic	30–69	9	60.5	3.87	63.8	3.05	
Iliac artery, arteriosclerotic	20–80	15	71.7	7.84	74.6	7.26	

1 Thoracic descending aorta.

A decrease in creatine content with age was observed for all the types of vascular tissue studied, the decline recorded for the thoracic descending aorta being statistically highly significant (table XVI). Of particular interest is the conspicuously lower creatine levels found in arteriosclerotic than in normal tissue portions (table XVII).

B. Animal Vascular Tissue

Using the same analytical procedure, assays have been made on 52 normal bovine arterial samples [KIRK, unpublished information]. The mean values listed in table XVIII indicate higher concentrations of total creatine in bovine than in human arterial tissues but the difference is not too pronounced when compared with those recorded for young persons.

Even higher creatine levels have been reported by DAEMERS-LAMBERT [1968, 1969] for bovine carotid, the calculated average content of 20 samples being 2.908 μM/g wet tissue. In muscular tissue of bovine mesenteric artery

Table XVIII. Mean total creatine concentrations of normal bovine arterial tissue. Values expressed as micromoles of total creatine per gram wet tissue and per gram tissue nitrogen

Vascular sample	Number	Wet tissue mean	SD distr.	Tissue nitrogen mean	SD distr.	Reference
Thoracic descending aorta	15	1.612	0.255	35.43	6.57	KIRK, unpublished information
Ascending aorta	9	1.264	0.216	27.78	4.75	
Abdominal aorta	10	1.571	0.298	35.62	7.08	
Pulmonary artery	15	1.519	0.293	41.16	8.75	
Coronary artery	3	1.029		29.99		
Abdominal aorta	4	1.000				DAEMERS-LAMBERT, 1964
Coronary artery	4	0.730				
Carotid	19	3.130				
Carotid[1]	20	2.908				DAEMERS-LAMBERT, 1968, 1969
Carotid[1]	7	2.622				DAEMERS-LAMBERT and ROLAND, 1967

1 Calculated by the present author.

(No. 6), LUNDHOLM and MOHME-LUNDHOLM [1962] found a mean total creatine concentration of 5.260 µM/g.

Analyses of rat aortic tissue performed by READ [1955] showed a total creatine content of 6.412 µM/g tissue (No. 16), which is much higher than in human and bovine aortas.

Because animal vascular samples can be made available for analysis immediately after sacrifice a few bovine creatine phosphate concentrations expressed as micromoles per gram wet tissue are presented. On the basis of the data reported by DAEMERS-LAMBERT in 1968 and 1969, an average creatine phosphate content of 0.69 in the carotid (No. 20) was calculated. In the mesenteric artery the creatine phosphate levels observed by LUNDHOLM and MOHME-LUNDHOLM [1962] and by BEVIZ and MOHME-LUNDHOLM [1965] were, respectively, 0.56 (No 6) and 0.20 (No. 10).

References

Beviz, A. and Mohme-Lundholm, E.: Influence of dihydroergotamine and adrenaline on the concentrations of glucose-6-phosphate, fructose-6-phosphate, adenosinetriphosphate, and creatine phosphate in bovine mesenteric artery. Acta physiol. scand. *65:* 289–291 (1965).

Chappell, J. B. and Perry, S. V.: Creatine phosphokinase. Assay and application for microdetermination of adenine nucleotides. Biochem. J. *57:* 421–427 (1954).

Daemers-Lambert, C.: Action du chlorure de potassium sur le métabolisme des esters phosphorés et le tonus du muscle artériel (carotide de bovidé). Angiologica, Basel *1:* 249–274 (1964).

Daemers-Lambert, C.: Dissociation par le fluorodinitrobezène des effets ATP-asiques, métabolique et contractile, lies a l'augmentation de la concentration en potassium extracellulaire, dans le muscle lisse artériel (carotide de bovidé). Angiologica, Basel *5:* 293–309 (1968).

Daemers-Lambert, C.: Action du fluorodinitrobenzène sur le métabolisme phosphoré du muscle lisse artériel pendant la stimulation eléctrique (carotide de bovidé). Angiologica, Basel *6:* 1–12 (1969).

Daemers-Lambert, C. et Roland, J.: Métabolisme des esters phosphorés pendant le developpement et le maintien de la tension phasique du muscle lisse artériel (carotides de bovidé). Angiologica, Basel *4:* 69–87 (1967).

Eggleton, P.; Eksden, S. R., and Gough, N.: The estimation of creatine and of diacetyl. Biochem. J. *37:* 526–529 (1943).

Kirk, J. E.: Variation with age in the creatine phosphokinase activity of human aortic tissue. J. Geront. *17:* 369–372 (1962).

Kirk, J. E.: Chemistry of the vascular wall of middle-sized arteries; in Orbison and Smith The peripheral blood vessels, pp. 45–72 (Williams & Wilkins, Baltimore 1963).

Kirk, J. E.: Enzyme activities of human inferior vena cava. Clin. Chem. *10:* 306–308 (1964).

Kirk, J. E.: Creatine phosphokinase; in Kirk Enzymes of the arterial wall, pp. 174–180 (Academic Press, New York 1969).

Lundholm, L. and Mohme-Lundholm, E.: The effects of adrenaline and glucose on the content of high-energy phosphate esters in substrate-depleted vascular smooth muscle. Acta physiol. scand. *56:* 130–139 (1962).

Pilsum, J. F. van; Martin, R. P.; Kito, E., and Hess, J.: Determination of creatine, creatinine, arginine, guanidinoacetic acid, guanidine and methylguanidine in biological fluids. J. biol. Chem. *222:* 225–235 (1956).

Read, W. O.: Mechanism of action of hypertensin. Amer. J. Physiol. *182:* 545–547 (1955).

Sutton, D. C.; Sutton, G. C.; Hinkens, G. F.; Buckingham, W. B., and Rondinelli, R.: Studies on whole blood and muscle creatine levels. Effect of systemic and local anoxia, of cardiac failure and compensation, and of alpha-tocopherol administration. Amer. Heart J. *47:* 67–77 (1954).

VII. Cytochrome c

Cytochrome c is an iron porphyrin protein which functions as a significant agent in oxidation-reduction reactions. It is a hydrogen carrier and has a fundamental role in cell respiration because in combination with cytochrome c oxidase (EC ferrocytochrome c: oxygen oxidoreductase; 1.9.3.1) it forms a most important respiratory enzyme system. Cytochrome c reductase (EC reduced-NAD: (acceptor) oxidoreductase; 1.6.99.3) mediates the transfer of electrons from NADH to oxidized cytochrome c, which is thus reduced; the activity of this enzyme in vascular tissue has been reported [KIRK, 1962].

The presence of cytochrome c in human arterial tissue was first demonstrated [KIRK, 1959] by its isolation from the thoracic descending aorta. Attempts have subsequently been made to determine its concentration in human aortic tissue [KIRK, unpublished information].

A. Human Vascular Tissue

1. Analytical Procedure

Because of the low cytochrome c content in arterial tissue it was considered preferable to perform each analysis on 25–50 g of tissue; this required using pooled samples.

The extraction of cytochrome c from the aortic tissue was carried out essentially as described by POTTER and DUBOIS [1942] and the isolation of cytochrome c from the extract was done by a modification of CARRUTHERS' method [1947]. The preparation of a 15% aqueous homogenate was made with a Kontes Dual homogenizer. All homogenate portions were transferred to a 600-ml beaker and the pH was adjusted to 3.5 by addition of 3% trichloroacetic acid solution. The mixture was then stirred for 90 min with a magnetic stirrer at room temperature, after which the beaker was covered with parafilm and placed in a deep-freeze refrigerator for 12 h. Following removal from the deep-freeze the sample in the beaker was thawed at room temperature and afterward stirred for 30 min with the magnetic stirrer.

The homogenate was then transferred to graduated centrifuge tubes and centrifuged for 10 min at 3,000 r. p. m. The volumes of observed precipitates were listed. The supernatants were added to a 600-ml beaker and pH adjusted to 7.0 by dropwise addition of a 0.5 N NaOH solution; the exact total volume was noted. The neutralized sample was left standing at room temperature for 10 min and thereafter centrifuged as described above, the resulting precipitate amounts again being recorded.

The supernatants were transferred to a flask to be used for isolation of cytochrome c. This was accomplished by letting the solution pass slowly through an aluminum oxide column (Baker and Adamson's aluminum oxide 1 D). The cytochrome c was eluted by repeated application of 3 ml 2 N NaOH. The eluate was neutralized and its volume then accurately measured.

The cytochrome c determination was done in a Beckman DU spectrophotometer at 550 and 535 nm after reducing the cytochrome to ferrocytochrome c by addition of a dithionite crystal to the cuvette.

To check the efficiency of the whole procedure as applied to tissue assay, measurements were made on two human heart specimens which showed 131.8 and 156.0 µg/g wet tissue. This is in good agreement with the findings of BIÖRCK [1956].

2. Results

The mean value obtained from 10 analyses of pooled large human aortic samples was 1.63 µg cytochrome c/g wet tissue (SD distr. 0.43). This concentration is slightly higher than that reported by ROSENTHAL and DRABKIN [1943] for human blood (1.3 µg/g).

References

BIÖRCK, G.: The content of cytochrome c in human heart and skeletal muscle. Acta med. scand. *154:* 305–316 (1956).

CARRUTHERS, C.: Polarographic determination of cytochrome c. J. biol. Chem. *171:* 641–651 (1947).

KIRK, J. E.: Enzyme activities of human arterial tissue. Ann. N. Y. Acad. Sci. *72:* 1006–1015 (1959).

KIRK, J. E.: The diaphorase and cytochrome c reductase activities of arterial tissue in individuals of various ages. J. Geront. *17:* 276–280 (1962).

POTTER, V. R. and DUBOIS, K. P.: The quantitative determination of cytochrome c. J. biol. Chem. *142:* 417–426 (1942).

ROSENTHAL, O. and DRABKIN, D. L.: Spectrophotometric studies. XI. The direct microspectrophotometric determination of cytochrome c. J. biol. Chem. *149:* 437–450 (1943).

VIII. Glutathione

Glutathione is a tripeptide, a combination of glutamic acid, cysteine and glycine. The functional group in the molecule is the thiol group. Many enzymes are -SH enzymes, only active in the thiol state, and it has been suggested that a very important biological function of glutathione is to keep these enzymes in the reduced form. It has been shown by ABDULLA and ADAMS [1970] that glutathione greatly increases the activity of lecithin: cholesterol transacylase in human aortic tissue; this is a sulfhydryl-dependent enzyme.

GSH is a specific coenzyme for glyoxalase I (EC S-lactoylglutathione methylglyoxal-lyase [isomerizing]; 4. 4. 1. 5) which catalyzes this reaction:

$$\text{methylglyoxal} + \text{GSH} \rightleftarrows \text{S-lactoylglutathione}.$$

This action of glutathione takes place by the addition of its -SH group across the C-C double bond of the enol form of methylglyoxal, a typical lyase reaction. It is interesting that an unusually high glyoxalase I activity has been recorded for human arterial tissue [KIRK and KIRK, 1958; KIRK, 1960]. In the second step of the glyoxalase system S-lactoylglutathione is hydrolyzed by glyoxalase II (EC S-2-hydroxyacylglutathione hydrolase; 3. 1. 2. 6) to lactate and free glutathione.

Determinations of glutathione reductase activity (EC reduced-NAD[P]: GSSG oxidoreductase; 1. 6. 4. 2) in human [KIRK, 1965, 1969] and animal vascular tissue [KIRK, 1969] have also been reported.

A. Total Glutathione

1. Human Vascular Tissue

Assays of total glutathione concentrations in human vascular tissue have been made by WANG and KIRK [1960]. Because GSSG and GSH can be converted to each other in tissues it was considered advisable to determine total glutathione (GSH plus GSSG) content.

a) Analytical Procedure

The employed technique has been described in detail by WANG and KIRK [1960]. The procedure of DOHAN and WOODWARD [1939] was used for electrolytic reduction of GSSG; the total GSH was measured by amperometric titration. Since reduced ascorbic acid is also included in the titration, the total titration result was corrected for the small amount of reduced ascorbic acid present in the sample, this latter value being determined on another aliquot by the 2,6-dichlorophenolindophenol method. In agreement with the report by DOHAN and WOODWARD it was demonstrated by WANG and KIRK that the electrolytic procedure does not cause reduction of dehydroascorbic acid.

b) Results

The total glutathione values listed in table XIX show average concentrations of 0.22 and 0.26 mg/g wet tissue for normal thoracic descending aorta and pulmonary artery, respectively, the difference between the two arteries being statistically significant when the glutathione contents are expressed per gram tissue nitrogen (table XX). A higher tendency to increase in glutathione with age was displayed by aortic than by pulmonary artery tissue (table XXI). Comparison of normal and arteriosclerotic portions from the same aortic samples revealed significantly lower glutathione levels in fibrous-arteriosclerotic and slightly higher in lipid-arteriosclerotic tissue (table XXII).

In human blood glutathione is present mainly in its reduced form (GSH) in red blood cells. The mean value found by BERTOLINI [1966] by assay of samples from 123 normal persons aged 10–90 years was 0.6825 mg GSH/ml erythrocytes. This glutathione concentration in red blood cells is definitely higher than the total glutathione (GSH plus GSSG) values observed for human arterial tissues.

2. Animal Vascular Tissue

Analyses of glutathione in 7 normal guinea pig aortic specimens have been reported by POLIKARPOVA [1961]. This investigation showed a mean value of 0.229 mg total glutathione/g wet tissue (range 0.187–0.316) which is very close to that found by WANG and KIRK [1960] for the human aorta.

B. Reduced Glutathione

1. Human Vascular Tissue

A rather extensive study of GSH in human aortic tissue has been published by ITO [1969]. The assays were performed by amperometric titration, but it is not mentioned whether correction was made by measuring reduced ascorbic acid in the samples.

The mean glutathione values for thoracic descending aorta and abdominal aorta from persons of various ages are presented in table XXIII; it should be noted that in this table ITO [1969] does not make a distinction between normal and arteriosclerotic tissue. The scientist points out that in younger subjects the abdominal aorta tends to have higher glutathione levels than the thoracic aorta. It is interesting that, whereas there was no definite change with age in reduced glutathione content for the thoracic aorta, a significant decrease ($r = -0.41$) was observed for the abdominal aorta.

The influence of arteriosclerosis on aortic tissue GSH content is listed in table XXIV. These data show that in the thoracic descending aorta the concentration was lowest in grade 3; in the abdominal aorta the decrease corresponded to the severity of arteriosclerosis. It might have been advisable to distinguish between lipid- and fibrous-arteriosclerotic tissue.

A somewhat similar investigation has later been reported by TAKAGI [1970], but in this research GSH was measured only on the intima layer of the aorta and concentrations expressed per gram dry tissue. The results presented in table XXV show, according to TAKAGI, no certain differences between normal thoracic aortic tissue and the 3 grades of arteriosclerotic lesions. In contrast to this, GSH in the abdominal aorta decreased significantly in the moderate (grade 2) and severe (grade 3) arteriosclerotic groups as compared with the normal group.

2. Animal Vascular Tissue

The effect of cholesterol feeding on GSH in rabbit aortic tissue has been studied by AOYAMA *et al.* [1967]. In 5 normal rabbits the mean concentration was 0.535 mg/g dry tissue (SD distr. 0.370). After 4 weeks (No. 5) and 12 months (No. 3) of cholesterol feeding the observed average GSH values were, respectively, 0.289 mg/g dry tissue (SD distr. 0.246) and 0.279 mg/g dry tissue (SD distr. 0.164). This clearly shows a decrease of GSH in the aortic wall following cholesterol feeding. An increase in peroxide content of the aortic tissue was already noted after 4 weeks experiment. These data are also included in TAKAGI's publication [1970].

Table XIX. Mean glutathione contents of human vascular tissue. Values expressed as milligrams of total glutathione per gram wet tissue and per gram tissue nitrogen

Vascular sample	Age group, years	Number	Wet tissue		Tissue nitrogen		Reference
			mean	SD distr.	mean	SD distr.	
Aorta normal[1]	0–9	6	0.2143	0.0592	4.740	1.331	WANG and
	10–19	7	0.1840	0.1272	4.117	2.739	KIRK, 1960
	20–29	5	0.1376	0.0404	3.244	0.771	
	30–39	10	0.1747	0.1204	4.236	2.509	
	40–49	13	0.2111	0.1063	6.058	3.278	
	50–59	13	0.2136	0.0892	5.929	2.397	
	60–69	13	0.3112	0.1017	8.726	2.693	
	70–79	4	0.2108		6.948		
	0–79	71	0.2171	0.1035	5.815	3.007	
	20–79	58	0.2214	0.1058	6.132	3.106	
Aorta, lipid-arteriosclerotic[1]	15	1	0.2810		6.779		
	30–39	2	0.0895		2.340		
	40–49	8	0.2651	0.0839	8.342	3.361	
	50–59	6	0.2672	0.1212	8.486	4.242	
	60–69	6	0.2828	0.0932	8.903	3.130	
	70–79	3	0.3090		10.083		
	15–79	26	0.2618	0.1062	8.184	3.483	
Aorta, fibrous-arteriosclerotic[1]	40–79	9	0.1449	0.1038	4.322	3.413	
Pulmonary artery, normal	0–9	1	0.3420		8.597		
	10–19	2	0.2821		7.790		
	20–29	1	0.1552		5.061		
	30–39	2	0.2450		7.075		
	40–49	6	0.2140	0.1213	6.623	4.102	
	50–59	6	0.2548	0.0583	7.583	1.732	
	60–69	9	0.3087	0.1142	9.288	3.766	
	70–79	2	0.1945		6.415		
	0–79	29	0.2597	0.1123	7.760	3.258	
	20–79	26	0.2548	0.1119	7.726	3.316	

1 Thoracic descending aorta.

Table XX. Mean glutathione concentrations of pulmonary artery samples expressed in percent of contents of normal aortic tissue from the same subjects

Vascular sample	Age group, years	Number	Wet tissue %	t of difference	Tissue nitrogen %	t of difference	Reference
Pulmonary artery	0–49	9	130.5	1.53	141.1	2.39	WANG and KIRK, 1960
	50–79	17	112.5	1.28	122.2	2.32	
	0–79	26	118.0	2.02	127.8	3.29	

Table XXI. Coefficients of correlation between age and tissue glutathione concentrations

Vascular sample	Age group, years	Number	Wet tissue r	t	Tissue nitrogen r	t	Reference
Aorta normal[1]	0–79	71	+0.28	2.42	+0.44	4.10	WANG and KIRK, 1960
	20–79	58	+0.40	3.26	+0.48	4.09	
Aorta, lipid-arteriosclerotic[1]	15–79	26	+0.32	1.68	+0.40	2.14	
Pulmonary artery, normal	0–79	29	+0.08	0.42	+0.16	0.85	
	20–79	26	+0.25	1.28	+0.24	1.22	

[1] Thoracic descending aorta.

Table XXII. Mean glutathione concentrations of arteriosclerotic tissue expressed in percent of contents of normal tissue portions from the same arterial samples

Vascular sample	Age group, years	Number	Wet tissue %	t of difference	Tissue nitrogen %	t of difference	Reference
Aorta, lipid-arteriosclerotic[1]	15–49	11	114.9	0.55	126.2	1.04	WANG and KIRK, 1960
	50–79	14	103.3	0.53	111.0	1.51	
	15–79	25	107.2	0.74	116.3	1.76	
Aorta, fibrous-arteriosclerotic[1]	40–79	9	70.7	3.40	81.1	2.12	

[1] Thoracic descending aorta.

Table XXIII. Mean concentrations of GSH in human aortic tissue[1]. Values expressed as milligrams of GSH per gram wet intima plus media tissue

Age group, years	Thoracic descending aorta			Abdominal aorta			Reference
	number	mean	SD distr.	number	mean	SD distr.	
11–20	6	0.228	0.128	6	0.302	0.204	ITO, 1969
21–40	5	0.229	0.132	3	0.275	0.206	
41–50	8	0.225	0.196	6	0.294	0.144	
51–60	13	0.232	0.114	10	0.264	0.160	
61–70	17	0.198	0.168	9	0.158	0.158	
71–80	5	0.232	0.036	2	0.028		

1 In this table distinction is not made between normal and arteriosclerotic tissue.

Table XXIV. Mean concentrations of GSH in normal and arteriosclerotic aortic samples from persons over 40 years of age

Vascular sample	Thoracic descending aorta			Abdominal aorta			Reference
	number	mean	SD distr.	number	mean	SD distr.	
Normal	5	0.199	0.150	4	0.280	0.144	ITO, 1969
Grade 1	11	0.225	0.160	7	0.270	0.158	
Grade 2	15	0.234	0.134	8	0.234	0.120	
Grade 3	5	0.140	0.162	4	0.111	0.126	

Values expressed as milligrams of reduced glutathione per gram wet intima plus media tissue. Severity of arteriosclerosis listed as grades 1, 2 and 3.

Table XXV. Mean concentrations of GSH in normal and arteriosclerotic aortic samples from persons aged 43–76 years

Vascular sample	Thoracic descending aorta			Abdominal aorta			Reference
	number	mean	SD distr.	number	mean	SD distr.	
Normal	4	0.213	0.156	4	0.289	0.126	TAKAGI, 1970
Grade 1	13	0.203	0.184	8	0.245	0.170	
Grade 2	16	0.219	0.148	9	0.210	0.160	
Grade 3	2	0.197	0.064	5	0.165	0.172	

Values expressed as milligrams of reduced glutathione per gram dry intima tissue. Severity of arteriosclerosis listed as grades 1, 2 and 3.

References

ABDULLA, Y. H. and ADAMS, C. W. M.: Sulphydryl dependence of arterial lecithin. Cholesterol transacylase. Atherosclerosis *12:* 319–320 (1970).

AOYAMA, S.; OHTA, H.; FUJISHIRO, N.; KAWASHIMA, S.; IWAHASHI, H.; TAKAGI, Y.; HASEGAWA, M.; MIYAZU, T., and ITO, T.: Fatty acid peroxides and antioxidants in rabbit tissue. Jap. Heart J. *8:* 142–147 (1967).

BERTOLINI, A. M.: Aging in red cells; in SHOCK Perspectives in experimental gerontology, pp. 156–168 (Thomas, Springfield 1966).

DOHAN, J. S. and WOODWARD, G. E.: Electrolytic reduction and determination of oxidized glutathione. J. biol. Chem. *129:* 393–403 (1939).

ITO, T.: Tocopherol, non-protein SH and metals in the human aorta. Jap. Circulat. J. (Engl. Ed.) *33:* 25–36 (1969).

KIRK, J. E.: The glyoxalase I activity of arterial tissue in individuals of various ages. J. Geront. *15:* 139–141 (1960).

KIRK, J. E.: The glutamic dehydrogenase and glutathione reductase activities of arterial tissue in individuals of various ages. J. Geront. *20:* 357–362 (1965).

KIRK, J. E.: Glutathione reductase; in KIRK Enzymes of the arterial wall, pp. 90–95 (Academic Press, New York 1969).

KIRK, J. E. and KIRK, T. E.: The glyoxalase I activity of human arterial tissue. J. Lab. clin. Med. *52:* 828–829 (1958).

POLIKARPOVA, L. I.: Glutathione contents in guinea pig blood vessels during irradiation (in Russian). Radiobiol. USSR *1:* 715–718 (1961).

TAKAGI, Y.: SH compounds and atherosclerosis. Nagoya J. med. Sci. *32:* 281–302 (1970).

WANG, I. and KIRK, J. E.: The total glutathione content of arterial tissue in individuals of various ages. J. Geront *15:* 35–37 (1960).

IX. Lipoic Acid

The significance of lipoic acid (thioctic acid; 6,8-dithiooctanoic acid) in carbohydrate metabolism has been established. It is an essential cofactor for the oxidative decarboxylation of pyruvic acid and α-ketoglutaric acid catalyzed by pyruvate dehydrogenase (EC pyruvate: lipoate oxidoreductase acceptor-acetylating; 1. 2. 4. 1) and α-ketoglutarate dehydrogenase (EC 2-oxoglutarate: lipoate oxidoreductase acceptor-acylating; 1. 2. 4. 2). This occurs in the presence of thiamine pyrophosphate. In these cases the reduction of lipoic acid is accompanied by the transfer of an acetyl group from the substrate through thiamine to the lipoate, giving 6-S-acetylhydrolipoate and 6-S-succinylhydrolipoate, respectively. The acetyl group is then transferred to CoA.

The reduced lipoic acid is reoxidized by lipoamide dehydrogenase (EC reduced-NAD: lipoamide oxidoreductase; 1. 6. 4. 3) coupled with the reduction of NAD. MASSEY [1960] has identified lipoamide dehydrogenase with the enzyme previously known as diaphorase; the activity of diaphorase in human vascular tissue has been studied by the present author [KIRK, 1962, 1963 a, b, 1964).

The presented data about the biological role of lipoic acid clearly show that it is an acyl-generating, an acyl-transferring and a hydrogen-transferring factor.

A. Human Vascular Tissue

Determinations of lipoic acid concentrations in human vascular tissue have been performed by KHEIM and KIRK [unpublished information].

1. Analytical Procedure

For determination of lipoic acid a modification of the technique described by STOKSTAD et al. [1956] was used. 100 mg of vascular tissue was homogenized in redistilled water, making a final volume of 10 ml.

a) Release of Protein-Bound Lipoic Acid

Because lipoic acid occurs in tissue largely in association with proteins it is necessary to release it for microbiological assay; this can be accomplished using the proteolytic trypsin enzyme. To the 10 ml vascular homogenate 10 ml 0.025% trypsin solution was added; this reagent was prepared by dissolving 25 mg purified trypsin (Sigma Chemical Co., St. Louis, Mo.) in 100 ml redistilled water, the pH being adjusted to 8.0.

A few drops of toluene were then added to the trypsin-containing homogenate which was poured into a screw-top tube. Following this, the tube was incubated for 24 h at 37^0 C. After the incubation period the screw top was replaced with a loose-fitting propylene top and the sample was autoclaved for 10 min at 15 lb pressure and 120^0 C to stop the enzyme activity and remove the toluene. The tube was then cooled and centrifuged, after which the supernatant was diluted with redistilled water to a volume of 50 ml. It was thereafter filtered through Whatman No. 42 filter paper, 1 ml of this filtrate corresponding to 2 mg vascular tissue.

b) Microbiological Assay of Lipoic Acid

For microbiological assay of lipoic acid the *Streptococcus faecalis* (ATCC 8043) method was used. Determinations were done in triplicate on both 1 and 2 ml of the described filtrate representing 2 and 4 mg tissue, respectively. The filtrate samples were pipetted into $18 \boxtimes 150$ mm disposable tubes which had been washed with an alcohol-HCL mixture and rinsed very well with distilled water. The volume of each tube was then made up to 5 ml with redistilled water after which 5 ml thioctic (lipoic) acid medium solution (Difco Labs., Inc., Detroit, Mich.) was added and the tubes covered with propylene cups (stoppers). They were thereafter autoclaved for 5 min at 15 lb pressure and 120^0 C and subsequently cooled quickly.

The inoculum was grown 18–22 h on 10 ml *Lactobacilli* broth AOAC (Difco Labs., Inc., Detroit, Mich.) supplemented with 10 mg papain digest of liver residue. At the end of this period the inoculum was centrifuged and washed with saline; the rinsed inoculum was then diluted 1:1000 with saline to make it available for use in microbiological assay of lipoic acid.

One drop of this prepared *S. faecalis* inoculum was put into each of the test tubes which were then incubated for 16 h at 37^0 C. After the incubation, 2 drops of a 10-percent formaldehyde solution were added to stop the growth of *S. faecalis*. The growth was measured turbidimetrically at 625 nm in a Beckman DU spectrophotometer. To acquire the most accurate results the tubes were shaken in a Vortex Genie Mixer before the samples were subjected to spectrophotometric assay.

With each vascular tissue analysis a blank was run and a reference curve was constructed. These were conducted in the same way as described above except that in the blank 10 ml redistilled water was used instead of 10 ml homogenate and for preparation of the standard curve solutions containing 0.4, 0.6, 0.8 and 1.2 ng of lipoic acid were employed.

2. Results

The observed lipoic acid concentrations for various types of human vascular tissue are presented in table XXVI. The mean value of 0.0993 µg/g wet tissue listed for normal thoracic descending aorta is somewhat higher than that reported by GUEDES et al. [1965] for 19 human serum samples (range 0.0152–0.0506 µg/ml serum; mean 0.0320 µg/ml serum, SD distr. 0.0236).

As seen from table XXVII statistically higher lipoic acid levels were found in the pulmonary artery and normal coronary artery than in thoracic descending aortic tissue from the same persons, whereas lower content was displayed by the vena cava inferior. No great change in lipoic acid with age occurred in the vascular samples studied except in lipid-arteriosclerotic coronary artery tissue for which a notable increase was recorded (table XXVIII).

Comparison of lipoic acid concentrations in arteriosclerotic and normal portions of the same arterial samples always showed higher content in the pathological tissue (table XXIX); the especially high values observed for lipid-arteriosclerotic aorta and coronary artery may be associated with the fact that lipoic acid is a fat-soluble compound.

Using the same analytical procedure, determinations have later been made of lipoic acid levels in ascending and abdominal aortic samples obtained from adult subjects. When expressed as micrograms per gram wet tissue the following mean lipoic acid concentrations were acquired: ascending aorta normal (No. 10) 0.0856 (SD distr. 0.0166); ascending aorta arteriosclerotic (No 7) 0.0901 (SD distr. 0.0202) abdominal aorta normal (No. 4) 0.1195, and abdominal aorta arteriosclerotic (No. 6) 0.1154 (SD distr. 0.0378).

Table XXVI. Mean lipoic acid concentrations of human vascular tissue. Values expressed as micrograms of lipoic acid per gram wet tissue and per gram tissue nitrogen

Vascular sample	Age group, years	Number	Wet tissue mean	SD distr.	Tissue nitrogen mean	SD distr.	Reference
Aorta normal[1]	15–19	6	0.0975	0.0268	2.617	0.751	KHEIM and
	20–29	7	0.1037	0.0168	2.719	0.448	KIRK, un-
	30–39	7	0.0992	0.0347	2.656	0.781	published
	40–49	7	0.1110	0.0485	3.051	1.136	information
	50–59	8	0.0910	0.0221	2.595	0.472	
	60–69	3	0.0840		2.420		
	70–87	3	0.1033		3.210		
	15–87	41	0.0993	0.0294	2.740	0.728	
Aorta, lipid-arteriosclerotic[1]	17–19	3	0.1323		3.640		
	20–29	4	0.1402		3.680		
	30–39	4	0.1158		3.438		
	40–49	4	0.1362		4.215		
	50–59	4	0.1208		4.345		
	60–69	3	0.1143		3.776		
	70–87	3	0.1450		4.677		
	17–87	25	0.1291	0.0314	3.959	1.118	
Aorta, fibrous-arteriosclerotic[1]	47–77	7	0.1109	0.0212	3.263	0.482	
Pulmonary artery, normal	15–19	4	0.1145		3.273		
	20–29	4	0.1015		3.083		
	30–39	2	0.0755		2.505		
	40–49	7	0.1162	0.0379	3.448	1.141	
	50–59	7	0.1117	0.0324	3.291	0.797	
	60–69	3	0.1150		3.663		
	70–87	3	0.1247		3.690		
	15–87	30	0.1110	0.0298	3.322	0.825	
Coronary artery, normal	17–19	5	0.1360	0.0308	4.104	0.888	
	20–29	3	0.1207		4.037		
	30–39	3	0.1060		3.467		
	40–49	4	0.1525		4.782		
	50–59	4	0.1073		3.580		
	60–77	3	0.1256		4.513		
	17–77	22	0.1261	0.0305	4.090	0.975	

1 Thoracic descending aorta.

Table XXVI (continued)

Vascular sample	Age group, years	Number	Wet tissue mean	SD distr.	Tissue nitrogen mean	SD distr.	Reference
Coronary artery, lipid-arteriosclerotic	19–29	3	0.1453		4.746		
	30–39	3	0.1327		5.163		
	40–49	3	0.1826		6.660		
	50–59	6	0.1500	0.0206	5.735	1.114	
	60–69	3	0.1647		5.870		
	70–87	4	0.1898		7.179		
	19–87	22	0.1607	0.0313	5.930	1.253	
Vena cava inferior	15–19	5	0.0746	0.0218	2.204	0.648	
	20–29	4	0.0605		1.855		
	30–39	3	0.0623		1.817		
	40–49	2	0.0645		1.830		
	50–59	4	0.0765		2.347		
	60–69	2	0.0705		2.025		
	70–87	2	0.0850		2.565		
	15–87	22	0.0704	0.0172	2.096	0.592	

Table XXVII. Mean lipoic acid concentrations of various types of normal vascular samples expressed in percent of contents of normal thoracic descending aortic tissue from the same subjects

Vascular sample	Age group, years	Number	Wet tissue %	t of difference	Tissue nitrogen %	t of difference	Reference
Pulmonary artery, normal	15–87	29	115.8	2.63	122.0	4.38	KHEIM and KIRK, unpublished information
Coronary artery, normal	17–77	20	136.0	4.97	157.5	6.63	
Vena cava inferior	15–87	22	73.6	4.76	80.0	3.57	

Table XXVIII. Coefficients of correlation between age and tissue lipoic acid concentrations

Vascular sample	Age group, years	Number	Wet tissue		Tissue nitrogen		Reference
			r	t	r	t	
Aorta normal[1]	15–87	41	—0.07	0.43	+0.08	0.50	Kheim and Kirk, unpublished information
Aorta, lipid-arteriosclerotic[1]	17–87	25	0.00	0.00	+0.25	1.24	
Pulmonary artery, normal	15–87	30	+0.19	1.03	+0.20	1.08	
Coronary artery, normal	17–77	22	—0.08	0.37	+0.12	0.54	
Coronary artery, lipid-arteriosclerotic	19–87	22	+0.47	2.43	+0.52	2.70	
Vena cava inferior	15–87	22	+0.21	0.99	+0.19	0.88	

1 Thoracic descending aorta.

Table XXIX. Mean lipoic acid concentrations of arteriosclerotic tissue expressed in percent of contents of normal tissue portions from the same arterial samples

Vascular sample	Age group, years	Number	Wet tissue		Tissue nitrogen		Reference
			%	t of difference	%	t of difference	
Aorta, lipid-arteriosclerotic[1]	17–87	25	142.3	7.96	154.4	7.85	Kheim and Kirk, unpublished information
Aorta, fibrous-arteriosclerotic[1]	47–77	7	119.2	3.64	121.0	3.53	
Coronary artery, lipid-arteriosclerotic	19–77	11	132.1	4.55	144.8	4.62	

1 Thoracic descending aorta.

References

GUEDES, M. F.; SANTOS MOTA, J. M., and ABREU, M. L.: Some technical problems of the thioctic acid microbiological assay in human serum. Med. Pharmacol. exp. *13:* 1–6 (1965).

KIRK, J. E.: The diaphorase and cytochrome c reductase activities of arterial tissue in individuals of various ages. J. Geront. *17:* 276–280 (1962).

KIRK, J. E.: Chemistry of the vascular wall of middle-sized arteries; in ORBISON and SMITH The peripheral blood vessels, pp. 45–72 (Williams & Wilkins, Baltimore 1963a).

KIRK, J. E.: A procedure for quantitative determination of the diaphorase activity of connective tissue. Clin. Chem. *9:* 776–779 (1963b).

KIRK, J. E.: Enzyme activities of human inferior vena cava. Clin Chem. *10:* 306–308 (1964).

MASSEY, V.: The identity of diaphorase and lipoyl dehydrogenase. Biochim. biophys. Acta *37:* 314–322 (1960).

STOKSTAD, E. L. R.; SEAMAN, G. R.; DAVIS, R. J., and HUTNER, S. H.: Assay of thioctic acid. Meth. biochem. Anal. *3:* 23–47 (1956).

X. Nucleotides and Nucleic Acids

A. Nucleotides

The nucleotides are extremely important factors in tissue metabolism. ATP has energy-rich phosphate stored as pyrophosphate bond. ATP has many biological functions; it provides the energy for muscular contraction and participates in protein biosynthesis. GTP is also required for protein synthesis; the reactions by which amino acids are activated and allowed to combine successively with the ribosomes are catalyzed by GTP-dependent enzymes. CTP is involved in biosynthesis of lecithin.

The biological carriers of phosphate are nucleoside diphosphates which act as cofactors in transphosphorylation processes. The main biological phosphate carrier is ADP. The enzymes which catalyze the transfer reactions to and from these diphosphates are the kinases (EC subgroups 2. 7. 1, 2. 7. 2, 2. 7. 3 and 2. 7. 4). CDP, GDP and UDP function as carriers in glycosyl transfer. The compound termed 'uridine coenzyme' is UDPgluc.

The pyridine nucleotides, NAD and NADP contain the nicotinic acid amide moiety derived from the vitamin nicotinic acid. The NAD and NADP coenzymes which operate as hydrogen and electron transfer agents by virtue of reversible oxidation and reduction play a vital role in metabolism.

The pentose sugars of nucleotides and nucleic acids are known to arise from the hexose monophosphate shunt (pentose phosphate pathway; direct oxidative shunt). The presence of the hexose monophosphate shunt in human arterial tissue was first demonstrated by the author [KIRK, 1958; KIRK et al., 1959].

Arterial tissue contains enzymes which change the contents of individual nucleotides. Rather high activities of ATPases (EC ATP phosphohydrolase, 3. 6. 1. 3; and ATP pyrophosphohydrolase, 3. 6. 1,8), 5'-nucleotidase (EC 5'-ribonucleotide phosphohydrolase; 3. 1. 3. 5) and myokinase (EC ATP: AMP phosphotransferase; 2. 7. 4. 3) in human vascular tissue have been published [KIRK, 1959 a, b, 1969]. Analyses performed immediately after samples have been acquired are therefore the most reliable. So far, most nucleotide measurements have been done on animal arterial tissue. Because of the great metabolic significance of the nucleotides, in the future such determinations should be made on human vascular specimens obtained at surgical operations.

1. Human Vascular Tissue

In the present author's department assays have been made of the pyridine nucleotide [CHANG et al., 1955] and the flavin nucleotide [SCHAUS et al., 1955] concentrations in human aortic tissue. The flavin nucleotide values were reported in the riboflavin chapter of a previous monograph [KIRK, 1973]. The pyridine nucleotide results will be described here.

a) Analytical Procedure

The intima-media of normal thoracic descending aortic tissue was homogenized with 9 vol of chilled 0.1 M phosphate buffer (pH 7.4) making a 10-percent homogenate. During the homogenization the tube of the grinder was kept immersed in ice water.

To determine the phosphopyridine nucleotide in aortic tissue the fluorometric methyl ethyl ketone alkali condensation procedure of BURCH [1952] was employed, using a standard solution of ^{1}N-methyl nicotinamide as reference. For each analysis 1.0 ml of homogenate was pipetted into a test tube, to which was immediately added 0.05 ml of the ceric sulfate-nicotinamide reagent. The content of the tube was carefully mixed, after which the protein was precipitated by addition of 0.10 ml 50% trichloroacetic acid solution, prepared from redistilled trichloroacetic acid crystals. The tube was cooled in ice water and centrifuged for 5 min.

50 mm^3 of the supernatant solution, representing 4.35 mg of wet tissue, were used for fluorometric determination. The fluorescence measurements were performed in a Farrand photoelectric fluorometer, using 3.0 ml cuvettes. The manganese chloride-methyl ketone solution, ^{1}N-methyl nicotinamide standard, and quinine sulfate reference solution were prepared fresh daily. Since the fluorometric method does not permit a distinction between NADP and NAD, the results are given as NAD concentrations.

b) Results

The NAD values found for various age groups are presented in table XXX. The mean concentrations recorded for adult persons: 23.85 µg/g wet tissue and 671.0 µg/g tissue nitrogen correspond to 0.0362 and 1.010 µM, respectively. The NAD level in aortic tissue is about half of that found by CHIEFFI and KIRK in human red blood cells [reported by KIRK, 1954].

Calculation of coefficients of correlation between age and NAD in aortic tissue revealed a moderate tendency to decrease with age. For the 0- to 84-year-old group (No. 49) the r values expressed on the basis of wet tissue weight and tissue nitrogen content were, respectively, -0.26 (t $= 1.85$) and -0.20 (t $= 1.40$). The corresponding coefficients of correlation for samples from 20- to 84-year-old persons (No. 44) were -0.26 (t $= 1.77$) and -0.25 (t $= 1.69$).

As pointed out by CHANG et al. [1955], the measured NAD concentrations must be considered to be lower than those present *in vivo* in the aortic wall because it was demonstrated that the tissue contains an enzyme which catalyzes the breakdown of NAD (EC NAD glycohydrolase; 3. 2. 2. 5); this enzyme is often termed 'NAD nucleosidase', 'NADase' and 'DPNase'. Assays of the activity of this enzyme made with the 10-percent homogenate prepared with phosphate buffer (pH 7.4) as described above and conducted at 37 °C showed an average disappearance of 19.4 µg NAD/g wet tissue/h. It is interesting to note that no change in total nicotinic acid content of the samples occured in the incubation tests.

2. Animal Vascular Tissue

During the last decade several studies have been reported about the nucleotides in animal arterial tissue. In 1959, BONFERONI et al., extracted acid-soluble nucleotides from calf aortic wall homogenates; by column chromatography followed by paper chromatography these investigators were able to identify 10 compounds: AMP, ADP, ATP, GMP, GTP, IMP, UMP, UDPgluc., UDP-hexosamine and UTP.

The presence of cyclic AMP (adenosine 3', 5'-monophosphate) in rat aortic tissue has been demonstrated by LARAIA and REDDY [1968] who found a concentration of 0.0006 µM/g wet tissue; it was further shown by these scientists that rat arterial tissue is capable of producing cyclic AMP. According to a report by BARTELSTONE et al. [1967] this nucleotide sometimes has a certain stimulating effect on the contraction of arterial tissue. In contrast to the findings by LARAIA and REDDY on rat aorta, HENRY et al. [1965] did not observe cyclic AMP in rabbit aortic tissue.

Another important research is that conducted by PICARD et al. [1962] who detected traces of PAPS in the rat aortic wall. This nucleotide sulfate is of metabolic significance because sulfate is transferred from PAPS to an acceptor such as chondroitin.

Most of the nucleotide analyses have been performed on bovine and rabbit arterial tissue. These quantitative values are presented in tables XXXI and XXXII. Measurements of the ATP content in dog renal artery tissue have been performed by BAYLIN et al. [1966]. In nonconstriction segments of this artery (No. 22) an average concentration of 0.1045 µM/g wet tissue (SD distr. 0.1686) was found. There was a much higher ATP level in constriction segments (No. 12), the mean value being 0.2270 µM/g wet tissue (SD distr. 0.2091).

The important studies by MANDEL et al. [1961] on bovine aortic samples have demonstrated a marked fall with age in tissue nucleotide concentrations (table XXXII). On the basis of their findings these scientists suggested that one of the metabolic aspects of aging in arterial tissue is related to a diminution of energy transfer, resulting in lowered rates of protein and enzyme synthesis; the decrease in adenosine nucleotides and especially in ATP certainly represents a reduction in metabolic energy reserve. Special assays conducted by MANDEL [1962 a, b] revealed lower rates of protein and RNA biosynthesis by the aorta of old than by young cows.

As seen from table XXXIII, the research by HENRY and GAUTHERON [1965] and HENRY et al. [1964, 1967 a, b, 1968] on rabbits receiving an atherogenic diet showed that a statistically significant decrease in ADP, ATP and NAD occurred in atherosclerotic tissue; the lower levels of ADP and ATP were not associated with the appearance of measurable AMP. The effect of the atherogenic diet on the biosynthesis of rabbit aortic tissue proteins has also been studied by HENRY [1968] and HENRY et al. [1968] who demonstrated that the rate of protein synthesis was distinctly reduced in aortic samples from rabbits with induced atherosclerosis as compared with normal aortic specimens.

B. Nucleic Acids

The nucleic acids, RNA and DNA are polynucleotides. It is widely agreed that recent research on nucleic acids has provided fascinating information in science. This includes especially the finding that DNA functions as the fundamental genetic substance. Many facts have also been acquired about the significance of the structure of the DNA chromosome. RNA is involved in protein synthesis; 3 types of RNA are present in cells: ribosomal-RNA, messenger-RNA and transfer-RNA. The synthesis of messenger-RNA takes place in the nucleus and is catalyzed by the RNA polymerase enzyme (EC nucleosidetriphosphate: RNA nucleotidyltransferase; 2. 7. 7. 6) with DNA as primer. In the future it can be expected that research on atherogenesis will be directed to studies on the nucleic acids in arterial cells of animal species with different susceptibility to atherosclerosis. In such investigations the nucleotide sequence in DNA and RNA will receive much attention.

The presence in bovine aortic tissue of an enzyme, ribonuclease which acts on RNA (EC ribonucleate pyrimidine–nucleotido–2'–transferase [cyclizing]; 2. 7. 7. 16) was demonstrated by GAMBLE et al. [1966, 1967] and by LEWIS et al. [1967]. These scientists [LEWIS and GAMBLE, 1969] have subsequently isolated two different ribonucleases from the bovine aorta, one of these enzymes, termed 'ribonuclease II' being similar to that previously found.

1. Human Vascular Tissue

Although there currently is a tendency to express enzyme activities of vascular tissue on the basis of its DNA content, data about nucleic acid concentrations in human arterial and venous samples are somewhat limited. The present author therefore decided to make a large number of such determinations [KIRK, unpublished information]. The results available from the literature will first be presented; a survey of such values is listed in table XXXIV.

The findings by EISENBERG et al. [1969 a, b] show that there is a marked fall in the DNA content of the aorta after the first decade; according to these authors the ratio of DNA to tissue wet weight remained constant between 10 and 97 years. Although PLATT and LUBOEINSKI [1969] reported a DNA concentration of aortic tissue from a newborn child of only 300 µg/g fresh weight they also found a definite DNA decrease with age. Until the age of 30 years this value declined to approximately 100 µg/g fresh weight; at this level the DNA content stayed fairly constant even in the high age groups. PLATT and LUBOEINSKI further observed that the DNA concentration at different stages of atherosclerosis varied considerably. The DNA and RNA contents of human vascular tissue acquired by analyses performed by the author will be described next.

a) Analytical Procedure

For determination of DNA and RNA nucleic acid concentrations the procedure of SCHNEIDER [1957] was used. The required diphenylamine and orcinol reagent solutions were prepared fresh daily from crystallized compounds. Standards and blanks were run with each set of analyses. Pure DNA and RNA were obtained from Sigma Chemical Co. (St. Louis, Mo.) and Calbiochem (San Diego, Calif.).

b) Results

The results of the 700 nucleic acid assays are presented in table XXXV. Several of the findings in adult persons are of interest. As seen from table XXXVI, significantly lower DNA and RNA concentrations were found in the pulmonary artery, coronary artery and especially in the vena cava inferior than in the thoracic descending aorta. Except in the coronary artery both nucleic acids showed a tendency to decrease with age (table XXXVII) and lower average contents were displayed by arteriosclerotic than by normal portions of the same arterial samples (table XXXVIII).

2. Animal Vascular Tissue

The nucleic acid concentrations for animal vascular tissue are listed in table XXXIX where the values are presented as milligrams of nucleic acids (DNA and RNA) and as micrograms of nucleic acid phosphorus (DNA-P and RNA-P) per gram tissue. In connection with this it shall be mentioned that according to SCHNEIDER [1945] the phosphorus content of DNA is 9.89% and of RNA 9.50%. In several of these investigations the ages of the studied animals are available.

EISENBERG *et al.* [1969 a, b] clearly demonstrated that in the rabbit the steepest fall in the DNA concentration of aortic tissue occurred between 1 and 3 months of age; during the same time the weight gain of the aorta was most pronounced (from an average of 263–513 mg).

According to EISENBERG *et al.* [1969 a, b] in the rat aorta the DNA level decreased progressively with age. As seen from the table, the greatest decline was observed between the ages of 1 and 1½ months whereas between 3 months and 2 years the variation was small. With regard to this finding, attention should be directed to the data by WORTMAN *et al.* [1966]. The assays by these scientists showed a more pronounced reduction in aortic tissue DNA concentration between newborn and 1-month-old rats, the reported DNA content per gram tissue at the age of 1 month being only 41% of that found for weanling rats; during this period of time the average weight of the aorta rose from 32.1 to 101.4 mg. It is interesting to note that in contrast to DNA no great change took place in RNA per gram tissue.

In the studies performed by MANDEL [1962 a, b] and PANTESCO *et al.*

[1962] assays of aortic samples from 2-years-old and 11- to 12-year-old bovines revealed only moderately lower DNA and RNA values in the tissue of the elderly animals. As also seen from the table, there was a slight increase with age in the contents of these nucleic acids in the wall of the vena cava inferior [PANTESCO et al., 1962].

Significantly higher DNA concentration was detected by NAKAMURA et al. [1965] in bovine coronary than aortic tissue, that of the coronary artery being close to that displayed by the myocardium; this is contrary to the finding by the present author for human coronary artery. The investigation by PANTESCO et al. [1962] showed that in bovines both the DNA and RNA levels were lower in the vena cava inferior tissue than in the aortic wall. PRIEST [1962, 1963] compared the DNA contents in the thoracic and abdominal aorta of adult rats; as seen from the table, these assays demonstrated distinctly higher DNA values for the thoracic aorta.

Reports are also available about the effect of atherogenic diet on aortic nucleic acid concentrations (table XL). In the research on monkeys [LEE et al., 1966] lower DNA and RNA occurred after 8 months of this diet whereas higher values appeared after 16-month feeding. In the experiments on rabbits no change in DNA content was observed by NERI SERNERI et al. [1962] but a definite decline was found by EISENBERG et al. [1969 a, b]; in connection with these controversial results it should be mentioned that NERI SERNERI et al. started the atherogenic diet with adult rabbits and EISENBERG et al. with 3-month-old animals. More studies are required in this field.

Table XXX. NAD content of normal human thoracic descending aortic tissue. Values expressed as micrograms of NAD per gram wet tissue and per gram tissue nitrogen

Age group, years	Number	Wet tissue		Tissue nitrogen		Reference
		mean	SD distr.	mean	SD distr.	
0– 9	5	28.22	10.82	723.6	245.6	CHANG
20–29	1	27.10		721.0		et al., 1955
30–39	4	26.30		712.3		
40–49	4	24.95		702.0		
50–59	10	28.45	10.83	798.8	283.2	
60–69	14	21.75	9.82	614.4	266.8	
70–84	11	20.86	8.28	596.5	210.2	
0–84	49	24.30	10.12	676.5	264.4	
20–84	44	23.85	10.13	671.0	265.2	

Table XXXI. Mean nucleotide concentrations of bovine arterial tissue[1]

Vascular sample	Number	Mean concentration, µM/g		Reference
Aorta, intima plus media				
AMP	8	0.33	DT	Nakamura et al.,
ADP	8	0.11	DT	1965
ATP	8	0.31	DT	
AMP	2	0.85	DT	Miyazaki and
ADP	2	0.85	DT	Nakamura, 1969
ATP	2	0.79	DT	
CMP	2	0.07	DT	
Hypoxanthine	2	0.41	DT	
NAD plus NADP	2	0.16	DT	
Abdominal aorta				
AMP	4	0.52	WT	Daemers-Lam-
ADP	4	0.33	WT	bert, 1964
ATP	4	0.37	WT	
GTP plus UTP	4	0.21	WT	
Coronary artery, intima plus media				
AMP	8	0.27	DT	Nakamura et al.,
ADP	8	0.54	DT	1965
ATP	8	0.81	DT	
AMP	3	1.74	DT	Miyazaki and
ADP	3	1.04	DT	Nakamura, 1969
ATP	3	0.14	DT	
CMP	3	0.20	DT	
Hypoxanthine	3	1.46	DT	
NAD plus NADP	3	0.24	DT	
Coronary artery				
AMP	4	0.34	WT	Daemers-Lam-
ADP	4	0.33	WT	bert, 1964
ATP	4	0.52	WT	
GTP plus UTP	4	0.20	WT	
Carotid				
AMP	19	0.82	WT	Daemers-Lam-
ADP	19	0.63	WT	bert, 1964
ATP	19	0.91	WT	
GTP plus UTP	19	0.38	WT	
AMP	8	0.690	WT	Daemers-Lam-
ADP	21	0.788	WT	bert, 1968, 1969
ATP	21	0.975	WT	
GTP plus UTP	21	0.449	WT	
Mesenteric artery				
AMP	6	1.24	WT	Lundholm and
ADP	6	0.92	WT	Mohme-Lund-
ATP	6	0.72	WT	holm, 1962
ATP	10	0.63	WT	Beviz and Mohme-Lundholm, 1965

[1] See also tables XXXII and XXXIII   DT = dry tissue, WT = wet tissue.

Table XXXII. Variation with age in nucleotide concentrations of bovine intima plus media aortic tissue[1]. Values expressed as micromoles per gram wet tissue

	Young calves	2-year-old cows	11- to 12-year-old cows
AMP	0.2900	0.4430	0.2102
ADP	0.0846	0.1012	0.0691
ATP	0.0134	0.0137	0.0060
CMP	0.0726	0.0622	0.0363
CDP-coenzymes	0.0299	0.0426	0.0361
GMP	0.0391	0.0498	0.0297
GDP	0.0012		
GTP	0.0016		
GDP-galactose	0.0182	0.0103	0.0087
GDP-mannose	0.0092	0.0081	0.0046
IMP	0.0072	0.0108	0.0060
NAD	0.0192	0.0188	0.0130
UMP	0.0363	0.0544	0.0306
UDP	0.0104	0.0104	0.0047
UDP-coenzymes	0.0737	0.0605	0.0418
UTP	0.0132		0.0088

1 References: young calves, MANDEL and KEMPF, 1961, KEMPF *et al.*, 1961; 2-year-old cows, KEMPF and MANDEL, 1961, MANDEL *et al.*, 1961, MANDEL, 1962 a, b; 11- to 12-year-old cows, KEMPF and MANDEL, 1961, KEMPF *et al.*, 1961, MANDEL *et al.*, 1961, MANDEL, 1962 a, b.

Table XXXIII. Effect of atherogenic diet on nucleotide concentrations in aortic intima plus media tissue of 3 different rabbit strains[1]

			Atherogenic diet		
	No.	Control	1 month	2 months	3½ months
ADP					
Strain I	10	0.246	0.218	0.150	0.129
Strain II	10	0.266			0.122
Strain III	10	0.311	0.168	0.056	0.056
ATP					
Strain I	10	0.147	0.110	0.094	0.043
Strain II	10	0.159			0.038
Strain III	10	0.192	0.111	0.029	0.019
NAD					
Strain I	10	0.109	0.109	0.080	0.058
Strain II	10	0.120			0.049
Strain III	10	0.140	0.063	0.038	0.038

1 No AMP or only traces of this compound were found.
Values expressed as micromoles per gram wet tissue (HENRY *et al.*, 1964; HENRY and GAUTHERON, 1965; HENRY *et al.*, 1967 a, b, 1968).

Table XXXIV. Mean concentrations of DNA in human arterial tissue

Vascular sample	Age group, years	Number	Mean concentration, mg/g		Reference
Ascending aorta, normal, intima plus media	0– 9	10	2.298	WT	Eisenberg et al., 1969
	10–19	7	1.542	WT	
	20–29	5	1.486	WT	
	30–39	4	1.445	WT	
	40–49	4	1.480	WT	
	50–59	4	1.612	WT	
	60–69	4	1.438	WT	
	70–79	3	1.453	WT	
	80–97	4	1.567	WT	
Thoracic aorta, normal, intima		10	1.284	WT[1]	Anastassiades and Denstedt, 1968
			5.0	DT	
Media		10	1.013	WT[1]	
			4.1	DT	
Thoracic aorta, atherosclerotic					
Intima		10	0.754	WT[1]	Anastassiades and Denstedt, 1968
			2.7	DT	
Media		10	1.172	WT[1]	
			4.2	DT	
Umbilical artery, intima plus media		7	1.16	WT	Rachmilewitz et al., 1967

1 Calculated by the present author.
DT = dry tissue, WT = wet tissue.

Table XXXV. Mean nucleic acid concentrations of human vascular tissue. Values expressed as milligrams of DNA and RNA per gram wet tissue (KIRK, unpublished information)

Vascular sample	Age group, years	Number	DNA mean	SD distr.	RNA mean	SD distr.
Aorta normal[1]	4– 9	3	1.223		1.630	
	10–19	4	1.160		1.365	
	20–29	10	1.270	0.264	1.505	0.273
	30–39	7	1.095	0.277	1.560	0.608
	40–49	7	1.168	0.265	1.660	0.239
	50–59	4	0.781		1.286	
	60–69	8	1.271	0.352	1.445	0.386
	70–75	4	0.886		1.382	
	4–75	47	1.142	0.324	1.493	0.384
	20–75	40	1.135	0.334	1.496	0.402
Aorta, lipid-arteriosclerotic[1]	10–19	3	1.518		1.608	
	20–29	10	1.157	0.137	1.331	0.349
	30–39	5	0.970	0.255	1.473	0.526
	40–49	5	1.064	0.180	1.499	0.229
	50–59	4	0.743		0.984	
	60–69	8	0.929	0.298	1.119	0.328
	70–75	4	0.876		1.207	
	10–75	39	1.032	0.317	1.300	0.381
	20–75	36	0.991	0.283	1.247	0.387
Aorta, fibrous-arteriosclerotic[1]	30–39	2	0.934		1.175	
	40–49	2	1.010		1.365	
	50–59	2	0.819		1.175	
	60–69	4	0.962		1.254	
	70–75	1	0.991		0.826	
	30–75	11	0.942	0.179	1.206	0.254
Ascending aorta, normal	4– 9	2	1.181		1.778	
	10–19	3	1.007		1.143	
	20–29	9	1.030	0.196	1.425	0.141
	30–39	7	1.045	0.255	1.360	0.203
	40–49	6	0.953	0.289	1.514	0.330
	50–59	4	0.800		1.366	
	60–69	4	0.972		1.381	
	70–75	4	0.991		1.191	
	4–75	39	0.993	0.241	1.389	0.323
	20–75	34	0.981	0.256	1.388	0.331

1 Thoracic descending aorta.

Table XXXV (continued)

Vascular sample	Age group, years	Number	DNA mean	SD distr.	RNA mean	SD distr.
Ascending aorta, arteriosclerotic	20–29	2	0.990		1.461	
	30–39	4	0.942		1.384	
	40–49	4	1.114		1.492	
	50–59	4	0.705		1.032	
	60–69	6	0.891	0.219	1.135	0.189
	70–75	4	0.781		1.397	
	20–75	24	0.892	0.273	1.290	0.374
Abdominal aorta, normal	4–9	3	1.245		1.418	
	10–19	4	1.122		1.254	
	20–29	7	1.323	0.129	1.580	0.187
	30–39	4	1.010		1.413	
	40–49	4	1.210		1.381	
	50–59	1	1.486		1.524	
	4–59	23	1.211	0.238	1.435	0.239
	20–59	16	1.228	0.249	1.484	0.270
Abdominal aorta, arteriosclerotic	20–29	5	1.295	0.188	1.499	0.389
	30–39	4	0.885		1.213	
	40–49	6	1.029	0.312	1.250	0.144
	50–59	4	0.600		0.811	
	60–69	4	0.882		1.238	
	70–75	4	0.678		1.016	
	20–75	27	0.920	0.354	1.193	0.353
Pulmonary artery, normal	4–9	2	1.371		1.270	
	10–19	3	1.301		1.122	
	20–29	10	1.026	0.211	1.308	0.470
	30–39	7	1.129	0.152	1.343	0.448
	40–49	7	1.106	0.271	1.243	0.178
	50–59	4	0.819		0.954	
	60–69	8	1.112	0.312	1.222	0.336
	70–75	4	0.915		1.255	
	4–75	45	1.076	0.276	1.238	0.341
	20–75	40	1.043	0.264	1.245	0.374
Coronary artery, normal	4–9	1	0.914		0.825	
	10–19	3	1.120		1.122	
	20–29	10	0.900	0.148	1.130	0.230
	30–39	6	0.807	0.173	1.046	0.194
	40–49	5	0.930	0.327	1.137	0.254

Table XXXV (continued)

Vascular sample	Age group, years	Number	DNA mean	SD distr.	RNA mean	SD distr.
	50–59	1	0.991		1.524	
	70–75	3	1.029		1.461	
	4–75	29	0.926	0.251	1.150	0.288
	20–75	25	0.903	0.244	1.167	0.303
Coronary artery, arteriosclerotic	30–39	4	0.924		1.952	
	40–49	4	0.971		1.397	
	50–59	4	0.676		1.000	
	60–69	7	0.899	0.237	1.108	0.227
	70–75	4	0.810		1.206	
	30–75	23	0.861	0.250	1.303	0.402
Vena cava inferior	4–9	3	0.533		0.762	
	10–19	4	0.527		0.619	
	20–29	10	0.454	0.124	0.775	0.176
	30–39	7	0.646	0.173	0.897	0.212
	40–49	7	0.566	0.158	0.738	0.118
	50–59	4	0.562		0.588	
	60–69	6	0.629	0.100	0.794	0.214
	70–75	2	0.343		0.635	
	4–75	43	0.545	0.158	0.754	0.178
	20–75	36	0.548	0.171	0.769	0.172

Table XXXVI. Mean nucleic acid concentrations of various types of normal vascular samples expressed in percent of contents of normal thoracic descending aortic tissue from the same subjects (KIRK, unpublished information)

Vascular sample	Age group, years	Number	DNA %	t of difference	RNA %	t of difference
Ascending aorta, normal	20–75	34	88.1	2.70	92.0	2.03
Abdominal aorta, normal	20–59	16	97.4	0.98	91.8	1.02
Pulmonary artery, normal	20–75	40	92.0	2.40	83.3	3.01
Coronary artery, normal	20–75	25	80.3	3.57	80.1	4.05
Vena cava inferior	20–75	36	49.1	11.26	51.6	10.40

Table XXXVII. Coefficients of correlation between age and tissue nucleic acid concentrations (KIRK, unpublished information)

Vascular sample	Age group, years	Number	DNA r	t	RNA r	t
Aorta normal[1]	20–75	40	—0.22	1.41	—0.17	1.09
Aorta, lipid-arteriosclerotic[1]	20–75	36	—0.39	2.48	—0.27	1.59
Ascending aorta, normal	20–75	34	—0.19	1.07	—0.06	0.34
Ascending aorta, arteriosclerotic	20–75	24	—0.26	1.22	—0.18	0.86
Abdominal aorta, normal	20–59	16	—0.06	0.22	—0.23	0.85
Abdominal aorta, arteriosclerotic	20–75	27	—0.44	2.42	—0.46	2.53
Pulmonary artery, normal	20–75	40	—0.10	0.62	—0.15	0.92
Coronary artery, normal	20–75	25	+0.14	0.66	+0.30	1.50
Coronary artery, arteriosclerotic	30–75	23	—0.15	0.69	—0.35	1.64
Vena cava inferior	20–75	36	+0.02	0.12	—0.18	1.04

[1] Thoracic descending aorta.

Table XXXVIII. Mean nucleic acid concentrations of arteriosclerotic tissue expressed in percent of contents of normal tissue portions from the same arterial samples (KIRK, unpublished information)

Vascular sample	Age group, years	Number	DNA %	t of difference	RNA %	t of difference
Aorta, lipid-arteriosclerotic[1]	20–75	36	86.5	4.61	86.1	3.85
Aorta, fibrous-arteriosclerotic[1]	30–75	11	79.3	4.03	78.6	3.43
Ascending aorta, arteriosclerotic	20–75	20	87.2	2.20	93.1	1.23
Abdominal aorta, arteriosclerotic	20–59	8	88.9	2.44	85.9	1.71
Coronary artery, arteriosclerotic	30–75	7	88.7	1.77	95.6	1.18

[1] Thoracic descending aorta.

Table XXXIX. Nucleic acid concentrations in animal vascular tissue

Animal and vascular sample	Age	Number	Mean concentration	Reference
Monkey aorta, Intima plus media	2–3 years	8	1.26 mg DNA/g wet tissue and 2.87 mg RNA/g wet tissue	Lee et al., 1966
Cow aorta Intima plus media	3–6 months	2	1.950 mg DNA/g wet tissue and 1.794 mg RNA/g wet tissue	Stein et al., 1970
	3–10 months	2	1.387 mg DNA/g wet tissue and 1.624 mg RNA/g wet tissue	
	young adult	3	270 µg DNA-P/g dry tissue	Nakamura et al., 1965
	2 years		50.1 µg DNA-P/g wet tissue and 79.8 µg RNA-P/g wet tissue	Mandel, 1962 a, b
	11–12 years		42.4 µg DNA-P/g wet tissue and 59.7 µg RNA-P/g wet tissue	
Intima plus media plus adventitia	2 years	6	46.7 µg DNA-P/g wet tissue and 78.9 µg RNA-P/g wet tissue	Pantesco et al., 1962
	11–12 years	6	37.9 µg DNA-P/g wet tissue and 73.8 µg RNA-P/g wet tissue	
Cow coronary artery Intima plus media	young adult	3	450 µg DNA-P/g dry tissue	Nakamura et al., 1965
Cow vena cava inferior Intima plus media plus adventitia	2 years	5	25.2 µg DNA-P/g wet tissue and 41.9 µg RNA-P/g wet tissue	Pantesco et al., 1962
	11–12 years	5	31.6 µg DNA-P/g wet tissue and 46.3 µg RNA-P/g wet tissue	
Dog aorta (whole aorta), Intima plus media		6	1.50 mg DNA/g wet tissue	Rachmilewitz et al., 1967
Rabbit aorta Intima plus media	adult	11	1.577 mg DNA/g wet tissue	Neri Serneri et al., 1962
Intima plus media plus adventitia	adult		63.7 µg DNA-P/g wet tissue and 84.5 µg RNA-P/g wet tissue	Mandel, 1962 a, b
Rabbit whole aorta Intima plus media	1 month	3	2.88 mg DNA/g wet tissue	Eisenberg et al., 1969
	3 months	3	2.07 mg DNA/g wet tissue	
	5 months	2	1.70 mg DNA/g wet tissue	
	6–9 months	6	1.76 mg DNA/g wer tissue	
	7–9 months	5	1.96 mg DNA/g wet tissue	
	12 months	4	1.77 mg DNA/g wet tissue	
	20–24 months	5	1.70 mg DNA/g wet tissue	

Table XXXIX (continued)

Animal and vascular sample	Age	Number	Mean concentration	Reference
	4–5 months	1	1.776 mg DNA/g wet tissue and 2.040 mg RNA/g wet tissue	STEIN et al., 1970
	adult	6	1.81 mg DNA/g wet tissue	RACHMILE-WITZ et al., 1967
Guinea pig aorta (whole aorta), Intima plus media	adult	3	1.91 mg DNA/g wet tissue	RACHMILE-WITZ et al., 1967
Rat thoracic aorta, Intima plus media	adult male		2.90 mg DNA/g wet tissue	PRIEST, 1962
	adult male	13	2.57 mg DNA/g wet tissue	PRIEST, 1963
Rat abdominal aorta, Intima plus media	adult male		2.10 mg DNA/g wet tissue	PRIEST, 1962
	adult male	14	1.86 mg DNA/g wet tissue	PRIEST, 1963
Rat whole aorta Intima plus media	newborn	20	4.29 mg DNA/g wet tissue[1] 21.7 mg DNA/g dry tissue and 3.24 mg RNA/g wet tissue[1] 16.4 mg RNA/g dry tissue	WORTMAN et al., 1966
	1 month	20	1.77 mg DNA/g wet tissue[1] 8.9 mg DNA/g dry tissue and 3.045 mg RNA/g wet tissue[1] 15.5 mg RNA/g dry tissue	
	1 month	3	3.12 mg DNA/g wet tissue	EISENBERG et al., 1969
	1½ months	3	2.22 mg DNA/g wet tissue	
	3 months	4	1.74 mg DNA/g wet tissue	
	6–12 months	5	1.80 mg DNA/g wet tissue	
	18–24 months	6	1.69 mg DNA/g wet tissue	
	adult male	20	401 µg DNA-P/g wet tissue	GERSCHENSON et al., 1962
	adult male	20	188 µg RNA-P/g wet tissue	
	adult female	22	435 µg DNA-P/g wet tissue	
	adult female	23	143 µg RNA-P/g wet tissue	
	adult male	10	383 µg DNA-P/g wet tissue	MALINOW, 1963
	adult male	10	181 µg RNA/P-g wet tissue	
	adult female	10	429 µg DNA-P/g wet tissue	
	adult female	12	144 µg RNA-P/g wet tissue	
	adult	2	1.62 mg DNA/g wet tissue	RACHMILE-WITZ et al., 1967
Intima plus media plus adventitia	adult		90.3 µg DNA-P/g wet tissue	MANDEL, 1962 a, b
	adult		105.4 µg RNA-P/g wet tissue	

[1] These nucleic acid concentrations per gram wet tissue were calculated by the present author.

Table XL. Effect of atherogenic diet on nucleic acid concentrations of animal aortic intima plus media tissue

Animal	Time of atherogenic diet	Number	Mean concentration	Reference
Monkey	control	8	1.26 mg DNA/g wet tissue and 2.87 mg RNA/g wet tissue	LEE et al., 1966
	8 months	8	0.99 mg DNA/g wet tissue and 2.23 mg RNA/g wet tissue	
	16 months	5	1.67 mg DNA/g wet tissue and 3.64 mg RNA/g wet tissue	
Rabbit	control	11	1.577 mg DNA/g wet tissue	NERI SERNERI et al., 1962
	1 month	4	1.597 mg DNA/g wet tissue	
	2 months	5	1.641 mg DNA/g wet tissue	
	3 months	8	1.534 mg DNA/g wet tissue	
	control	3	2.07 mg DNA/g wet tissue	EISENBERG et al., 1969
	8 weeks	2	1.87 mg DNA/g wet tissue	
	17–18 weeks	4	1.66 mg DNA/g wet tissue	
	23 weeks	4	1.45 mg DNA/g wet tissue	

References

ANASTASSIADES, T. and DENSTEDT, O. F.: Phosphoglucoisomerase activity in the atherosclerotic aorta. Canad. J. Biochem. 46: 671–675 (1968).

BARTELSTONE, H. J.; NASMYTH, P. A., and TELFORD, J. M.: The significance of adenosine cyclic 3', 5'-monophosphate for the contraction of smooth muscle. J. Physiol., Lond. 188: 159–176 (1967).

BAYLIN, G. J.; DEMARIA, W. J.; BAYLIN, S. B.; KRUEGER, R. P., and SANDERS, A. P.: ATP concentration and localization of sites of epinephrine induced renal artery constriction Proc. Soc. exp. Biol. Med. 122: 396–399 (1966).

BEVIZ, A. and MOHME-LUNDHOLM, E.: Influence of dihydroergotamine and adrenaline on the concentrations of glucose-6-phosphate, fructose-6-phosphate, adenosinetriphosphate and creatine phosphate in bovine mesenteric artery. Acta physiol. scand. 65: 289–291 (1965).

BONFERONI, B.; RONCHI, S. e ZAMBOTTI, V.: Nucleotidi acido solubili nella parete aortica di vitello. Rev. esp. Fisiol. 15: 311–318 (1959).

BURCH, H. B.: Microfluorometric measurement of phosphopyridine nucleotides in blood serum and cells. Fed. Proc. 11: 192–193 (1952).

CHANG, Y. O.; LAURSEN, T. J. S., and KIRK, J. E.: The total nicotinic acid and pyridine nucleotide content of human aortic tissue. J. Geront. 10: 165–169 (1955).

CHIEFFI, M. and KIRK, J. E.: DPN content of human blood; data presented in KIRK Blood and urine vitamin levels in the aged. Nutrition symposium series No. 9, pp. 73–94 (The National Vitamin Foundation, New York 1954).

DAEMERS-LAMBERT, C.: Action du chlorure de potassium sur le métabolisme des esters phosphorés et le tonus du muscle artériel (carotide de bovidé). Angiologica, Basel 1: 249–274 (1964).

DAEMERS-LAMBERT, C.: Dissociation par le fluorodinitrobenzène des effets ATP-asiques, métabolique et contractile, lies a l'augmentation de la concentration en potassium extracellulaire, dans le muscle lisse artériel (carotide de bovidé). Angiologica, Basel 5: 293–309 (1968).

DAEMERS-LAMBERT, C.: Action du fluorodinitrobenzène sur le métabolisme phosphoré du muscle lisse artériel pendant la stimulation electrique (carotide de bovidé). Angiologica, Basel 6: 1–12 (1969).

EISENBERG, S.; STEIN, Y., and STEIN, O.: Phospholipases in arterial tissue. III. Phosphatide acyl-hydrolase, lysophosphatide acyl-hydrolase and sphingomyelin choline phosphohydrolase in rat and rabbit aorta in different age groups. Biochim. biophys. Acta 176: 557–569 (1969a).

EISENBERG, S.; STEIN, Y., and STEIN, O.: Phospholipases in arterial tissue. IV. The role of phosphatide acyl hydrolase, lysophosphatide acyl hydrolase, and sphingomyelin choline phosphohydrolase in the regulation of phospholipid composition in the normal human aorta with age. J. clin. Invest. 48: 2320–2329 (1969b).

GAMBLE, W.; NAYAR, G. H., and KIERSKY, E. S.: Bovine aorta ribonuclease. Fed. Proc. 25: 789 (1966).

GAMBLE, W.; NAYAR, G. H., and KIERSKY, E. S.: Bovine aorta ribonuclease. Biochim. biophys. Acta 145: 260–271 (1967).

GERSCHENSON, L.; MALINOV, M. R.; LACUARA, J. L., and MOGUILEVSKY, H. C.: Changes in the aortic concentration of nucleic acids induced by gonadectomy in rats. J. Atheroscler. Res. 2: 365–372 (1962).

HENRY, J. C.: Effets d'un régime athérogène sur la biosynthèse des protéines du tissue aortique. C. R. Acad. Sc. 266: 1449–1450 (1968).

HENRY, J. C. et GAUTHERON, D.: Répartition des nucléotides et des dérivés phosphorylés du tissu aortique de lapins normaux et soumis a un régime hypercholestérolique. Bull. Soc. Chim. biol. 47: 213–222 (1965).

HENRY, J. C.; GAUTHERON, D. et GRAS, J.: Fractions phosphorylées et nucléotides adényliques de l'aorte chez le lapin normal et soumis a un régime athéroslcérogène. C. R. Acad. Sci. 258: 3581–3583 (1964).

HENRY, J. C.; GAUTHERON, D.; GRAS, J. et FREY, J.: Variations des nucléotides adényliques et des fractions phosphorylées du tissu aortique sous l'influence de l'adrénaline. Bull. Soc. Chim. biol. 47: 1941–1950 (1965).

HENRY, J. C.; GRAS, J,; FREY, J.; MEROUZE, P.; ROUSSET, D. et COUCHAT, M. C.: Evolution des nucléotides riches en energie du tissu aortique et du myocarde chez le lapin soumis a un régime athérogène. G. Arterioscler. 5: 21–26 (1967a).

HENRY, J. C.; GRAS, J.; FREY, J.; MEROUZE, P.; ROUSSET, D. et COUCHAT, M. C.: Evolution des nucléotides riches en energie du tissu aortique et du myocarde chez le lapin soumis a un régime athérogène. Rev. Atheroscler. 9: 30–34 (1967b).

HENRY, J. C.; GRAS, J. et PERRIN, A.: Activités métaboliques comparées de la paroi artérielle normale et athérosclereuse. Rev. lyon. Med. 17: 753–762 (1968).

KEMPF, E.; FONTAINE, R. et MANDEL, P.: Etude comparée des nucléotides libres: adenyliques et uridyliques, des aortes de bovidés jeunes et agés. C. R. Soc. Biol. 155: 623–625 (1961).

Kempf, E. et Mandel, P.: Les nucléotides libres des aortes des bovidés adultes et agés. C. R. Acad. Sci. *253:* 2155–2157 (1961).

Kirk, J. E.: The presence of the direct oxidative shunt in human arterial tissue. Circulation *18:* 487 (1958).

Kirk, J. E.: The adenylpyrophosphatase, inorganic pyrophosphatase, and phosphomonoesterase activities of human arterial tissue in individuals of various ages. J. Geront. *14:* 181–188 (1959 a).

Kirk, J. E.: The 5-nucleotidase activity of human arterial tissue in individuals of various ages. J. Geront. *14:* 288–291 (1959b).

Kirk, J. E.: Adenosinetriphosphatases; in Kirk Enzymes of the arterial wall, pp. 362–384 (Academic Press, New York 1969).

Kirk, J. E.: Vitamin contents of arterial tissue. Monographs on atherosclerosis, Vol. 3 (Karger, Basel 1973).

Kirk, J. E.; Wang, I., and Brandstrup, N.: The glucose-6-phosphate and 6-phosphogluconate dehydrogenase activities of arterial tissue in individuals of various ages. J. Geront. *14:* 25–31 (1959).

Laraia, P. J. and Reddy, W. J.: Adenosine 3', 5'-monophosphate in aorta. Levels and regulation. Circulation *38:* suppl. 6 P. 122 (1968).

Lee, K. T.; Jones, R.; Kim, D. N.; Florentin, R.; Coulston, F., and Thomas, W. A.: Studies related to protein synthesis in experimental animals fed atherogenic diets. IV. ^{14}C-glycine incorporation into proteins of aortas and electron microscopy of non-necrotic atherosclerotic lesions of monkeys fed atherogenic diets for 8 to 16 months. Exp. molec. Path.: suppl. 3, pp. 108–123 (1966).

Lewis, R. A. and Gamble, W.: Purification and characterization of bovine aorta ribonucleases. Biochem. J. *115:* 95–101 (1969).

Lewis, R. A.; Kiersky, E. S., ans Gamble, W.: Specificity of bovine aorta ribonuclease. Fed. Proc. *26:* 840 (1967).

Lundholm, L. and Mohme-Lundholm, E.: The effects of adrenaline and glucose on the content of high-energy phosphate esters in substrate-depleted vascular smooth muscle. Acta physiol. scand. *56:* 130–139 (1962).

Malinow, M. R.: Modification du métabolisme artériel par les hormones. Rev. Atheroscler. *5:* 41–48 (1963).

Mandel, P.: Métabolisme de la paroi artérielle et ses modifications au cours du vieillissement. Int. Congr. Angeiologie, 1961. In: Metabolismus parietis vasorum, pp. 25–32 (State Medical Publishing House, Prague 1962a).

Mandel, P.: Role du métabolisme de la paroi artérielle dans la génèse de l'athérome. Arch. Sci. med. *113:* 223–236 (1962b).

Mandel, P. et Kempf, E.: Les nucléotides libres du tissu aortique. Biochim. biophys. Acta *51:* 184–186 (1961).

Mandel, P.; Pantesco, V.; Kempf, E. et Fontaine, R.: Métabolisme énergétique de la paroi artérielle au cours du vieillissement et athérosclerose. 23 rd Congr. Français de Médecine, pp. 228–235 (Masson, Paris 1961).

Miyazaki, T. and Nakamura, M.: Acid-soluble nucleotides analysis of bovine coronary artery and aorta. Jap. Heart J. *10:* 259–266 (1969).

Nakamura, M.; Miyazaki, T.; Sata, T., and Ishihara, Y.: Specificity of energy metabolism of coronary artery. Arzneimittelforsch. *15:* 1382–1388 (1965).

NERI SERNERI, G. G.; FRANCHI, F. e IGNESTI, C.: Studio di alcune attività enzimatiche della parete aortica del coniglio in corso di arteriopatia sperimentale da colesterolo. G. Geront. *10:* 1293–1309 (1962).

PANTESCO, V.; KEMPF, E.; MANDEL, P. et FONTAINE, R.: Études métaboliques comparées des parois artérielle et veineuse chez les bovidés. Leurs variations au cours du vieillissement. Path. Biol. *10:* 1301–1306 (1962).

PICARD, J.; GARDAIS, A. et LACCORD, M.: Présence de nucléotide sulfate dans la paroi de l'aorte. C. R. Acad. Sci. *255:* 2182–2183 (1962).

PLATT, D. and LUBOEINSKI, H. P.: The activities of glycosaminoglycan hydrolases of normal and atherosclerotic human aorta. Angiologica, Basel *6:* 19–31 (1969).

PRIEST, R. E.: Differences in metabolism of proximal and distal aorta. Circulation *26:* 667–668 (1962).

PRIEST, R. E.: Consumption of oxygen by thoracic and abdominal aorta of the rat. Amer. J. Physiol. *205:* 1200–1202 (1963).

RACHMILEWITZ, D.; EISENBERG, S.; STEIN, Y., and STEIN, O.: Phospholipases in arterial tissue. I. Sphingomyelin cholinephosphohydrolase activity in human, dog, guinea pig, rat and rabbit arteries. Biochim. biophys. Acta *144:* 624–632 (1967).

SCHAUS, R.; KIRK, J. E., and LAURSEN, T. J. S.: The riboflavin content of human aortic tissue. J. Geront. *10:* 170–177 (1955).

SCHNEIDER, W. C.: Phosphorus compounds in animal tissues. I. Extraction and estimation of deoxypentose nucleic acid and of pentose nucleic acid. J. biol. Chem. *161:* 293–303 (1945).

SCHNEIDER, W. C.: Determination of nucleic acids in tissues by pentose analysis; in COLOWICK and KAPLAN Methods in enzymology, vol. 3, pp. 680–684 (Academic Press, New York 1957).

STEIN, O.; RACHMILEWITZ, D.; EISENBERG, S., and STEIN, Y.: Aortic phospholipids. Israel J. med. Sci. *6:* 53–66 (1970).

WORTMAN, B.; LEE, K. T.; KIM, D. N.; DAOUD, A. S., and THOMAS, W. A.: Studies related to protein synthesis in experimental animals fed atherogenic diets. II. DNA, RNA and protein in aortas of rats fed atherogenic diets for 1 month. Exp. molec. Path.: suppl. 3, pp. 88–95 (1966).

XI. Ubiquinone

Ubiquinone (coenzyme Q) is a significant metabolic factor. It participates in electron transport and seems to occupy a key position between the flavoproteins and the cytochromes in the respiratory system of mitochondria. The compound of the ubiquinone group which is present in human tissues is ubiquinone 10 (ubiquinone [50]; coenzyme Q_{10}). It is a neutral lipid and is insoluble in water.

A. Human Vascular Tissue

An extensive study of ubiquinone concentrations in human vascular tissue has been completed by the present author [KIRK, unpublished information].

1. Analytical Procedure

One gram of vascular tissue was homogenized with 10 ml redistilled water and the homogenate transferred to a 50-ml amber extraction flask; for quantitative transfer the homogenizer was washed with small amounts of water which were poured into the extraction flask.

a) Saponification

Saponification of the homogenate was performed in the following way: 4 ml 2.5% pyrogallol solution and a boiling stone were added to the extraction flask which was then connected with a reflux condenser and heated for 3 min at 98° C. After this preheating of the sample 250 mg of crystallized potassium hydroxide pellets were added through the reflux condenser and heating continued for 15 min. The extraction flask was then cooled by placing it in tap water and subsequently in ice water.

b) Extraction of Ubiquinone

For extraction of ubiquinone from the saponified sample 3 amber separatory funnels of 125 ml capacity with glass-stopper were used; the glass-stoppers were greased with Nonaq Stopcock Grease (Fisher Scientific Co., St. Louis, Mo.). 25 ml ether (specially purified diethyl ether) was placed in each funnel. The content of the extraction flask was transferred to funnel No. 1; for quantitative transfer the extraction flask was washed with small amounts of water. The glass-stoppered funnel was then shaken for 3 min and afterwards left standing until the layers had separated. The lower aqueous section was then added to funnel No. 2 which was similarly shaken for 3 min and the subsequently separated aqueous layer transferred to funnel No. 3, the content of which was handled in the same way, and the resulting water portion was discarded.

c) Combination of Ether Extracts

The ether extracts from separatory funnels No. 2 and No. 3 were added to funnel No. 1. Funnels No. 2 and No. 3 were then washed with 5–10 ml ether, these ether washings also being transferred to funnel No. 1.

d) Washing of Combined Ether Extracts in Funnel No. 1

25 ml redistilled water was added to the combined ether extracts. The funnel was then shaken for 3 min, after which it was left standing until the ether and aqueous layers were clearly separated. The water was then run out of the lower part of the funnel and its pH measured. Washing with 25 ml water was made two more times; the out-running water from the third washing was usually no longer alkaline. In all tests washing was continued until all traces of alkali had disappeared.

e) Evaporation of the Combined and Washed Ether Extract

The ether sample was poured from funnel No. 1 into a beaker and repeated ether washings of the funnel added to the beaker. The ether extract in the beaker was then evaporated using a stream of nitrogen, after which the residue was dissolved in absolute alcohol, care being taken to rotate several times to insure that the residue would go completely into solution. Each alcoholic sample was transferred quantitatively to a graduated tube where the volume was adjusted to 5.0 ml.

f) Spectrophotometric Determination of Ubiquinone

The procedure used for ubiquinone determination was based on the colorimetric method described by REDALIEU et al. [1968b] in which the use of potassium hydroxide in the reaction between ubiquinone and ethylcyanoacetate has been replaced with benzyltrimethylammonium hydroxide; this gives nearly twice as strong blue color development as that obtained with the potassium hydroxide reagent.

For assay of normal arterial samples the alcoholic volumes in the graduated tubes were reduced from 5.0 to 2.0 ml with a stream of nitrogen; 1.5 ml of the concentrated solution was pipetted off and 0.75 ml ethylcyanoacetate and 0.75 ml 0.8 N bezyltrimethylammonium hydroxide in absolute alcohol were added. Immediately after these additions the sample was centrifuged for 2 min to eliminate any turbidity and then transferred to a cuvette in a Beckman DU spectrophotometer where optical density reading of the blue color was made at 630 nm against a reference cuvette containing absolute alcohol. These spectrophotometric readings were always done 3–4 min after addition of the reagents.

Analysis of arteriosclerotic and vena cava inferior samples was done in a similar way but using other volumes: the 5.0 ml of alcoholic tissue solution in the graduated tube were reduced to 3.0 ml, 2.5 ml were pipetted off and 1.25 ml ethylcyanoacetate and 1.25 ml benzyltrimethylammonium hydroxide added.

Reagent blanks for these two sets of volumes in which the 1.5 and 2.5 ml tissue solutions were replaced with 1.5 and 2.5 ml absolute alcohol were run 2 or 3 times daily. If the reagent blanks made in the morning gave a slight optical density reading, the benzyltrimethylammonium hydroxide solution was filtered through celite as recommended by REDALIEU *et al.* [1968b].

A standard curve was prepared using pure coenzyme Q_{10} obtained from the Calbiochem Co. (Los Angeles, Calif.).

g) Paper Chromatography of Ubiquinone

Ubiquinone was demonstrated chromatographically for each type of vascular tissue studied. This was accomplished by the ascending chromatography technique described by CRANE and DILLEY [1963]. In the procedure selected Whatman No. 3 MM paper was used. It was impregnated with silicone by immersing it in a 5-percent Dow-Corning Silicone Fluid No. 550 (Dow-Corning Corp., Midland, Mich.) in chloroform for several seconds, after which the paper was dried by exposing it to air at room temperature.

In each experiment a few residues of the concentrated alcoholic tissue solutions remaining after quantitative ubiquinone determinations were mixed for each type of vascular tissue; these mixtures usually contained about 3–8 µg of ubiquinone. After reduction to a very small volume by a stream of nitrogen it was applied to the silicone-treated paper. Five micrograms of pure coenzyme Q_{10} were placed on another paper. The fluid used for chromatography was the 4:1 propanol-water reagent recommended by CRANE and DILLEY [1963].

The chromatography period was 16–18 h. The equipment was not exposed to light and the air in the chromatography apparatus was replaced with hydrogen. For detection of ubiquinone on the paper, spraying with leucomethylene blue was made; this reagent was always prepared just before use.

2. Results

The mean ubiquinone concentrations of human vascular tissues presented in table XLI show an average content of 2.696 µg/g wet tissue for the normal thoracic descending aorta. This is about 3 times higher than the mean value of 0.83 µg/ml normal blood (range 0.40–1.80) observed by REDALIEU et al. [1968a] on the basis of 30 analyses. In connection with the 143 ubiquinone determinations of normal aortic tissue listed in table XLI it should be mentioned that FOLKERS et al. [1961], in a single analysis of an aorta from a 29-year-old person, found a coenzyme Q_{10} content of 0.6 mg/100 g tissue ($= 6$ µg/g tissue). Although this is higher than the mean values for 20- to 29- and 30- to 39-year-old age groups it is still within the range, especially if some lipid-arteriosclerotic tissue was included in that single assay.

As seen from table XLII, statistically higher ubiquinone concentrations were found in the pulmonary artery, normal coronary artery and inferior vena cava than in the thoracic descending aorta. A significant increase in ubiquinone with age was observed for the pulmonary artery, fibrous-arteriosclerotic thoracic aorta and arteriosclerotic abdominal aorta (table XLIII).

A comparison of ubiquinone contents in arteriosclerotic and normal tissue portions from the same arterial samples (table XLIV) showed notably higher values for lipid-arteriosclerotic tissue of the thoracic descending aorta and for arteriosclerotic portions of the ascending and abdominal aortas.

B. Animal Vascular Tissue

Using the same analytical procedure as described for human vascular tissue ubiquinone determinations were made on some bovine arterial samples [KIRK, unpublished information]. The mean concentrations observed for thoracic descending aorta (No. 5) were 0.936 µg/g wet tissue (SD distr. 0.212) and 23.64 µg/g tissue nitrogen (SD distr. 5.78). The corresponding values for the pulmonary artery (No. 9) were 1.152 (SD distr. 0.283) and 31.12 (SD distr. 8.06). These data indicate lower ubiquinone levels in bovine than in human arterial tissue. It is interesting to note that similar to the finding for human samples higher ubiquinone content was displayed by bovine pulmonary artery than by aortic tissue.

Ubiquinone in the bovine arteries was also identified by paper chromatography.

Table XLI. Mean ubiquinone concentrations of human vascular tissue. Values expressed as micrograms of ubiquinone per gram wet tissue and per gram tissue nitrogen

Vascular sample	Age group, years	Number	Wet tissue mean	SD distr.	Tissue nitrogen mean	SD distr.	Reference
Aorta normal[1]	15–19	11	1.827	0.515	48.36	15.98	KIRK, un-
	20–29	19	2.872	1.384	73.05	34.64	published
	30–39	12	3.110	1.612	83.02	43.86	informa-
	40–49	14	2.610	1.089	71.50	30.65	tion
	50–59	12	2.613	0.985	74.42	31.32	
	60–69	12	2.902	1.252	82.25	37.61	
	70–87	3	3.033		88.00		
	15–87	83	2.696	1.223	73.02	35.28	
Aorta, lipid-arteriosclerotic[1]	17–19	4	3.385		93.96		
	20–29	13	4.241	1.562	105.08	30.74	
	30–39	9	5.097	2.280	144.22	62.21	
	40–49	12	4.544	3.658	136.67	119.46	
	50–59	8	4.118	1.764	132.75	58.90	
	60–69	11	4.796	2.456	151.27	83.63	
	70–87	5	5.066	1.624	169.04	58.52	
	17–87	62	4.514	2.265	133.08	73.96	
Aorta, fibrous-arteriosclerotic[1]	39–49	6	2.548	0.849	71.17	26.46	
	50–59	4	3.057		93.96		
	60–69	2	5.024		157.52		
	70–79	3	5.343		162.67		
	39–79	15	3.573	1.484	107.07	48.16	
Ascending aorta, normal	16–19	3	2.380		62.67		
	20–29	5	1.938	0.363	51.44	11.72	
	30–39	8	3.117	1.345	83.75	35.26	

1 Thoracic descending aorta.

Table XLI (continued)

Vascular sample	Age group, years	Number	Wet tissue mean	SD distr.	Tissue nitrogen mean	SD distr.	Reference
	40–49	8	2.386	0.572	67.25	16.04	
	50–59	5	2.452	1.073	69.58	30.82	
	60–69	5	2.412	0.532	72.80	21.79	
	70–87	3	1.797		50.72		
	16–87	37	2.448	0.826	68.52	24.83	
Ascending aorta, arteriosclerotic	20–29	2	2.375		61.54		
	30–39	4	4.292		122.75		
	40–49	4	5.455		155.23		
	50–59	4	4.418		121.50		
	60–69	4	3.978		110.76		
	70–79	4	4.632		141.88		
	20–79	22	4.357	1.464	124.18	46.51	
Abdominal aorta, normal	15–19	9	2 555	1.183	76.33	36.74	
	20–29	2	1.986		61.52		
	30–39	3	2.823		83.66		
	40–49	6	3.062	1.897	95.17	60.33	
	50–65	3	2.714		91.04		
	15–65	23	2.694	1.128	82.83	37.59	
Abdominal aorta, arteriosclerotic	19–29	10	4.384	1.825	137.29	55.76	
	30–39	9	5.327	2.872	163.78	86.03	
	40–49	7	4.723	1.166	171.00	63.26	
	50–59	10	6.561	3.017	216.32	87.92	
	60–69	7	6.211	2.677	192.28	88.41	
	70–87	3	5.847		218.94		
	19–87	46	5.467	2.062	178.48	77.20	
Pulmonary artery, normal	15–19	10	2.566	1.462	72.64	38.99	
	20–29	12	2.551	0.931	76.75	35.08	
	30–39	11	3.855	1.206	113.00	38.60	
	40–49	16	3.568	1.469	107.06	52.71	
	50–59	12	3.387	1.164	101.92	36.52	
	60–69	13	3.507	1.079	105.98	37.94	
	70–87	5	3.936	0.531	116.20	15.81	
	15–87	79	3.312	1.196	98.54	42.07	
Coronary artery, normal	15–19	4	3.267		104.53		
	20–29	5	4.310	1.124	138.11	39.68	

Table XLI (continued)

Vascular sample	Age group, years	Number	Wet tissue mean	SD distr.	Tissue nitrogen mean	SD distr.	Reference
	30–39	5	5.536	1.536	186.74	44.43	
	40–49	4	5.568		174.52		
	50–59	4	4.768		146.09		
	60–65	3	4.053		127.96		
	15–65	25	4.633	1.425	148.32	50.19	
Coronary artery, arteriosclerotic	22–39	3	5.833		241.34		
	40–49	6	7.305	2.645	250.82	99.58	
	50–59	6	5.697	2.024	208.50	87.86	
	60–69	6	5.360	1.713	180.67	55.86	
	70–87	5	5.774	2.312	195.40	84.37	
	22–87	26	6.021	2.017	213.12	81.85	
Vena cava inferior	15–19	8	4.480	2.421	126.25	63.87	
	20–29	13	4.604	2.966	137.54	85.02	
	30–39	7	5.221	3.028	150.14	92.89	
	40–49	6	3.822	1.945	118.02	77.07	
	50–59	6	4.686	2.560	141.00	86.75	
	60–69	7	4.870	2.515	145.86	80.31	
	70–87	4	4.313		134.09		
	15–87	51	4.601	2.482	136.47	80.62	

Table XLII. Mean ubiquinone concentrations of various types of normal vascular samples expressed in percent of contents of normal thoracic descending aortic tissue from the same subjects

Vascular sample	Age group, years	Number	Wet tissue %	t of difference	Tissue nitrogen %	t of difference	Reference
Ascending aorta, normal	16–87	31	89.0	1.10	88.9	1.23	KIRK, unpublished information
Abdominal aorta, normal	15–65	21	104.3	0.91	118.6	1.17	
Pulmonary artery, normal	15–87	67	130.4	4.34	140.8	5.15	
Coronary artery, normal	15–65	22	166.6	3.81	192.1	4.45	
Vena cava inferior	15–87	41	193.8	5.72	199.5	5.65	

Table XLIII. Coefficients of correlation between age and tissue ubiquinone concentrations

Vascular sample	Age group, years	Number	Wet tissue		Tissue nitrogen		Reference
			r	t	r	t	
Aorta normal[1]	20–87	72	−0.03	0.25	+0.06	0.51	KIRK, unpublished informaiton
Aorta, lipid-arteriosclerotic[1]	20–87	58	+0.10	0.75	+0.25	1.94	
Aorta, fibrous-arteriosclerotic[1]	39–79	15	+0.59	2.68	+0.68	3.35	
Ascending aorta, normal	20–87	34	−0.12	0.66	−0.02	0.11	
Ascending aorta, arteriosclerotic	20–79	22	+0.17	0.76	+0.25	1.15	
Abdominal aorta, normal	20–65	14	+0.22	0.76	+0.29	1.08	
Abdominal aorta, arteriosclerotic	20–87	44	+0.30	2.04	+0.32	2.19	
Pulmonary artery, normal	20–87	69	+0.24	2.03	+0.24	2.03	
Coronary artery, normal	20–65	21	+0.04	0.17	+0.01	0.04	
Coronary artery, arteriosclerotic	22–87	26	−0.07	0.34	−0.16	0.78	
Vena cava inferior	20–87	43	−0.04	0.25	−0.01	0.06	

1 Thoracic descending aorta.

Table XLIV. Mean ubiquinone concentrations of arteriosclerotic tissue expressed in percent of contents of normal tissue portions from the same arterial samples

Vascular sample	Age group, years	Number	Wet tissue		Tissue nitrogen		Reference
			%	t of difference	%	t of difference	
Aorta, lipid-arteriosclerotic[1]	19–87	53	166.3	5.03	178.6	5.82	KIRK, unpublished information
Aorta, fibrous-arteriosclerotic[1]	39–79	14	123.7	1.24	129.4	1.30	
Ascending aorta, arteriosclerotic	29–79	12	172.1	3.11	182.7	3.76	
Abdominal aorta, arteriosclerotic	19–65	10	201.7	3.38	215.3	3.06	

1 Thoracic descending aorta.

References

CRANE, F. L. and DILLEY, R. A.: Determination of coenzyme Q (ubiquinone). Meth. biochem. Anal. *11:* 279–306 (1963).

FOLKERS, K.; SHUNK, C. H.; LINN, B. O.; TRENNER, N. R.; WOLF, D. E.; HOFFMAN, C. H.; PAGE, A. C., jr., and KONIUSZY, F. R.: Coenzyme Q. XXIII. Organic and biological studies; in WOLSTENHOLME and O'CONNER Ciba foundation symposium on quinones in electron transport, pp. 100–126 (Churchill, London 1961).

REDALIEU, E.; NILSSON, I. M.; FARLEY, T. M.; FOLKERS, K., and KONIUSZY, F. R.: Determination and levels of coenzyme Q_{10} in human blood. Analyt. Biochem. *23:* 132–140 (1968a).

READLIEU, E.; NILSSON, I. M.; NILSSON, J. L. G.; KJAER-PEDERSEN, D. I., and FOLKERS, K.: New procedure for assay and stability of coenzyme Q_{10} in human blood. Int. Z. Vitaminforsch. *38:* 345–354 (1968b).

Summary

This monograph contains the results of numerous coenzyme and metabolic factor analyses performed on human vascular tissue. The compounds studied generally displayed higher concentrations in normal aortic tissue than those reported for plasma or serum.

To facilitate acquisition about some of the assumed important findings obtained by determinations on human aortic tissue a few short summaries are presented here in 3 tables. The statistical significance of each of the reported values and the number of samples analyzed are listed in the tables of the various chapters. In the first 2 tables the selected data, except for DNA and RNA, are those calculated on the basis of tissue nitrogen content.

A comparison of coenzyme concentrations in normal, lipid-arteriosclerotic and fibrous-arteriosclerotic portions of the same arterial samples is considered to be of definite interest. The percentage values in the first survey table clearly show that some of the coenzymes (lipoic acid and ubiquinone) have markedly higher levels in lipid-arteriosclerotic than in normal tissue of the thoracic descending aorta.

Mean coenzyme and metabolic factor concentrations of arteriosclerotic tissue in human thoracic descending aorta expressed in percent of contents of normal tissue portions from the same aortic samples:

	Lipid-arteriosclerotic tissue, %	Fibrous-arteriosclerotic tissue, %
Carnitine (free carnitine)	101.8	88.9
Carnosine	110.7	106.9
CoA	102.6	64.5
Creatine (total)	67.3	67.0
Glutathione (total)	116.3	81.1
Lipoic acid	154.4	121.0
DNA	86.5	79.3
RNA	86.1	78.6
Ubiquinone	178.6	129.4

The increase with aging in the degree of arteriosclerosis is well known. A summary of coefficients of correlation between age and coenzyme contents of normal and arteriosclerotic human aortic tissue is, therefore, also presented. One information provided by this table is the difference between the coefficients of correlation for fibrous- and lipid-arteriosclerotic aortic tissue observed for some of the coenzymes studied.

Coefficients of correlation in adult persons between age and coenzyme concentrations of normal and arteriosclerotic thoracic descending aortic tissue:

	Normal aortic tissue, r	Lipid-arteriosclerotic tissue, r	Fibrous-arteriosclerotic tissue, r
Carnitine (free carnitine)	—0.06	+0.02	+0.04
Carnosine	+0.15	+0.32	+0.08
CoA	+0.21	+0.04	—0.24
Creatine (total)	—0.48	—0.20	—0.33
Glutathione (total)	+0.48	+0.40	
Lipoic acid	+0.08	+0.25	
DNA	—0.22	—0.39	
RNA	—0.17	—0.27	
Ubiquinone	+0.06	+0.25	+0.68

Because the pulmonary artery and inferior vena cava are less susceptible to pathological changes, it was considered advisable also to summarize the coenzyme levels of these blood vessels expressed in percent of concentrations observed for normal thoracic descending aortic tissue from the same persons.

Mean coenzyme concentrations of pulmonary artery and vena cava inferior expressed in percent of contents of normal thoracic descending aortic tissue from the same subjects:

Summery

	Pulmonary artery, %	Vena cava inferior, %
Carnitine (free carnitine)	115.4	54.5
Carnosine	97.9	174.6
CoA	127.4	130.1
Creatine (total)	140.6	55.4
Glutathione (total)	118.0	
Lipoic acid	115.8	73.6
DNA	92.0	49.1
RNA	83.3	51.6
Ubiquinone	130.4	193.8

This survey table shows that for many of the metabolic compounds investigated there is little similarity in the percentage values recorded for the pulmonary artery and the vena cava; differences in anatomic structure may to some extent account for it.

Subject Index

Acetylcarnitine 6
Adenosine diphosphate (ADP)
- bovine, aorta abdominal 63
- - aorta thoracic 58, 63, 64
- - carotid 63
- - coronary artery 58, 63
- - mesenteric artery 63
- rabbit, aorta atherosclerotic 59, 64
- - aorta normal 59, 64
Adenosine monophosphate (AMP)
- bovine, aorta abdominal 63
- - aorta thoracic 58, 63, 64
- - carotid 63
- - coronary artery 63
- - mesenteric artery 63
Adenosine 3′,5′-monophosphate (cyclic AMP)
- rat, aorta 58
Adenosine triphosphatases 56
Adenosine triphosphate (ATP)
- bovine, aorta abdominal 63
- - aorta thoracic 58, 59, 63, 64
- - carotid 63
- - coronary artery 63
- - mesenteric artery 63
- dog, renal artery 59
- rabbit, aorta atherosclerotic 59, 64
- - aorta normal 59, 64
β-Alanyl-histidine, *see* Carnosine
Animal vascular tissue, coenzyme contents of
- bovine, aorta abdominal 38, 63
- - aorta ascending 38
- - aorta thoracic 37, 38, 58, 59, 61, 63, 64, 70, 79
- - carotid 37, 38, 63
- - coronary artery 38, 62, 63, 70
- - mesenteric artery 38, 63

- - pulmonary artery 38, 80
- - vena cava inferior 62, 70
- chicken, aorta normal 26
- dog, aorta normal 70
- - renal artery 59
- guinea pig, aorta normal 26, 43, 71
- hamster, aorta normal 26
- monkey, aorta atherosclerotic 62, 72
- - aorta normal 62, 70, 72
- pigeon, aorta normal 26
- rabbit, aorta atherosclerotic 44, 59, 62, 64, 72
- - aorta normal 26, 44, 59, 61, 62, 64, 70, 72
- rat, aorta abdominal 62, 71
- - aorta thoracic 62, 71
- - whole aorta normal 26, 38, 58, 61, 62, 71
Atherosclerosis, definition of 1

Carnitine (free carnitine) 6–12
- analytical procedure 7
- biological function 6
- in human vascular tissue 7–10
Carnitine acetyltransferase 6
Carnosine 13–20
- analytical procedure 14
- biological function 13
- in human vascular tissue 14–20
Citrate condensing enzyme 21
Coefficients of correlation between age and coenzyme contents of human vascular tissue
- aorta abdominal arteriosclerotic 19, 36, 69, 83
- aorta abdominal normal 19, 27, 36, 69, 83

Subject Index

- aorta ascending arteriosclerotic 19, 36, 69, 83
- aorta ascending normal 19, 27, 36, 69, 83
- aorta thoracic descending fibrous-arteriosclerotic 11, 19, 27, 36, 83
- aorta thoracic descending lipid-arteriosclerotic 11, 19, 27, 36, 46, 54, 69, 83
- aorta thoracic descending normal 11, 19, 27, 36, 46, 54, 58, 69, 83
- coronary artery arteriosclerotic 11, 19, 27, 36, 54, 69, 83
- coronary artery normal 11, 19, 27, 36, 54, 69, 83
- iliac artery arteriosclerotic 19, 36
- iliac artery normal 19, 36
- iliac vein 19, 36
- pulmonary artery normal 11, 19, 27, 36, 46, 54, 69, 83
- vena cava inferior 11, 19, 27, 36, 54, 69, 83

Coenzyme A 21–29
- analytical procedure 21–23
- biological function 21
- in animal vascular tissue 27, 28
- in human vascular tissue 21–28

Coenzyme Q, see Ubiquinone

Comparison of coenzyme concentrations in arteriosclerotic and normal portions of the same human arterial samples
- aorta abdominal 20, 28, 37, 69, 83
- aorta ascending 20, 37, 69, 83
- aorta thoracic descending, about fibrous-arteriosclerotic tissue 12, 28, 37, 46, 54, 69, 83
- aorta thoracic descending, about lipid-arteriosclerotic tissue 12, 20, 28, 37, 46, 54, 69, 83
- coronary artery 12, 20, 28, 37, 54, 69
- iliac artery 20, 37

Comparison of coenzyme concentrations in various types of normal vascular tissue with contents in normal thoracic descending aortic tissue from the same subjects
- aorta abdominal normal 11, 18, 26, 35, 68, 82
- aorta ascending normal 11, 18, 26, 35, 68, 82

- coronary artery normal 11, 18, 26, 35, 53, 68, 82
- iliac artery normal 11, 18, 35
- iliac vein 11, 19, 35
- pulmonary artery normal 11, 18, 26, 35, 46, 53, 68, 82
- vena cava inferior 11, 18, 26, 35, 53, 68, 82

Creatine (total creatine) 30–39
- analytical procedure 30, 31
- biological function 30
- in animal vascular tissue 37, 38
- in human vascular tissue 30–37

Creatine phosphate in animal vascular tissue 38

Creatine phosphokinase 30

Cyclic AMP, see Adenosine 3′,5′-monophosphate

Cytidine diphosphate (CDP)
- bovine, aorta thoracic 64

Cytidine diphosphate coenzymes
- bovine, aorta thoracic 64

Cytidine monophosphate (CMP)
- bovine, aorta 63, 64
- – coronary artery 63

Cytochrome c
- analytical procedure 40, 41
- biological function 40
- in human vascular tissue 40, 41

Cytochrome c oxidase 40

Cytochrome c reductase 40

Deoxyribonucleic acid (DNA) in animal vascular tissue
- bovine, aorta 61, 70
- – coronary artery 62, 70
- – vena cava inferior 62, 70
- dog, aorta 70
- guinea pig, aorta 71
- monkey, aorta atherosclerotic 62, 72
- – aorta normal 62, 70, 72
- rabbit, aorta atherosclerotic 62, 72
- – aorta normal 61, 62, 71, 72
- rat, aorta abdominal 62, 71
- – aorta thoracic 62, 71
- – whole aorta 61, 71

Deoxyribonucleic acid (DNA) in human vascular tissue
- aorta abdominal arteriosclerotic 67
- aorta abdominal normal 67
- aorta ascending arteriosclerotic 67
- aorta ascending normal 60, 65, 66
- aorta thoracic descending fibrous-arteriosclerotic 60, 61, 66
- aorta thoracic descending lipid-arteriosclerotic 60, 61, 65, 66
- aorta thoracic descending normal 60, 61, 65, 66
- coronary artery arteriosclerotic 68
- coronary artery normal 61, 67
- pulmonary artery normal 61, 67
- umbilical artery 65
- vena cava inferior 61, 68

Diaphorase, see Lipoamide dehydrogenase
Direct oxidative shunt, see Hexose monophosphate shunt
6,8-Dithiooctanoic acid, see Lipoic acid
DPN, see Nicotinamide-adenine dinucleotide (NAD)

Effect of coenzymes and some metabolic factors on experimental atherosclerosis in animals 3, 4, 5
Effect of experimental atherosclerosis on coenzyme contents in animal vascular tissue 28, 44, 59, 62, 64, 72

Glutathione 42–48
- biological function 42
- reduced glutathione 44, 47
- - in animal vascular tissue 44
- - in human vascular tissue 44, 47
- total glutathione 42, 45, 46
- - analytical procedure 43
- - in animal vascular tissue 43
- - in human vascular tissue 42, 43, 45, 46

Glutathione reductase 42
Glyoxalase I 42
Guanosine diphosphate (GDP)
- bovine, aorta 64
Guanosine monophosphate (GMP)
- bovine, aorta 58, 64
Guanosine triphosphate (GTP)
- bovine, aorta abdominal 63
- - aorta thoracic 58, 63, 64
- - carotid 63
- - coronary artery 63

Hexose monophosphate shunt 56
Human vascular tissue, coenzyme concentrations of
- aorta abdominal arteriosclerotic 9, 16, 25, 32, 44, 47, 51, 66, 79, 81
- aorta abdominal normal 9, 14, 16, 23, 24, 32, 44, 47, 51, 67, 81
- aorta ascending arteriosclerotic 9, 15, 32, 51, 67, 79, 81
- aorta ascending normal 9, 15, 23, 24, 32, 51, 60, 65, 66, 80
- aorta thoracic descending fibrous-arteriosclerotic 8, 15, 23, 24, 31, 43, 45, 52, 66, 79, 80
- aorta thoracic descending lipid-arteriosclerotic 8, 15, 23, 24, 31, 43, 45, 47, 51, 52, 60, 61, 65, 66, 79, 80
- aorta thoracic descending normal 7, 8, 14, 15, 24, 31, 34, 41, 43–45, 47, 51, 52, 57, 60–62, 65, 66, 79, 80
- coronary artery arteriosclerotic 10, 17, 23, 25, 33, 53, 68, 82
- coronary artery normal 8, 9, 14, 16, 23, 25, 33, 51, 52, 61, 67, 79, 81
- iliac artery arteriosclerotic 10, 17, 34
- iliac artery normal 10, 14, 17, 33
- iliac vein 8, 10, 14, 18, 34, 35
- pulmonary artery normal 8, 9, 16, 23, 25, 33, 35, 43, 45, 51, 52, 61, 67, 79, 81
- umbilical artery 65
- vena cava inferior 8, 10, 14, 17, 23, 25, 34, 35, 51, 53, 61, 68, 79, 82

β-Hydroxy γ-butyrobetaine, see Carnitine

Inosine monophosphate (IMP)
- bovine, aorta 58, 64

α-Ketoglutarate dehydrogenase 49
Kinases 56

Lecithin: cholesterol transacylase 42
Lipmann unit 23
Lipoamide dehydrogenase 49
Lipoic acid 49–55
- analytical procedure 49–51
- biological function 49
- in human vascular tissue 49–54

α-Methylguanidine, see Creatine
Myokinase 56
Myosin ATPase 13

NAD nucleosidase 58
Nicotinamide-adenine dinucleotide + nicotinamide-adenine dinucleotidephosphate (NAD + NADP)
in animal vascular tissue
- bovine, aorta 63, 64
- - coronary artery 63
- rabbit, aorta atherosclerotic 64
- - aorta normal 64
in human vascular tissue
- aorta thoracic descending normal 57, 58, 62
Nucleic acids 59–72
- analytical procedure 60, 61
- biological function 59, 60
- in animal vascular tissue 61, 62, 70–72
- in human vascular tissue 60, 61, 65–69
5'-Nucleotidase 56
Nucleotides 56–59, 62–64, 72–75
- analytical procedure 57
- biological function 56
- in animal vascular tissue 58, 59, 63, 64
- in human vascular tissue 57, 58, 62

Pantothenic acid 21–23
PAPS, see 3'Phosphoadenosine-5'-phosphosulfate
Pentose phosphate pathway, see Hexose monophosphate shunt
3'Phosphoadenosine-5'-phosphosulfate (PAPS)
- rat, aorta 58
3-Phosphoglyceraldehyde dehydrogenase 13

Protein biosynthesis 56, 59
Pyridine nucleotides, see Nicotinamide-adenine dinucleotide + nicotinamide-adenine dinucleotidephosphate (NAD + NADP)
Pyruvate dehydrogenase 13, 49

Ribonuclease 60
Ribonucleic acid (RNA) biosynthesis 59
Ribonucleic acid (RNA) in animal vascular tissue
- bovine, aorta 61, 62, 70
- - vena cava inferior 62, 70
- monkey, aorta atherosclerotic 62, 72
- - aorta normal 62, 70, 72
- rabbit, aorta normal 70, 71
- rat, whole aorta 61, 71
Ribonucleic acid (RNA) in human vascular tissue
- aorta abdominal arteriosclerotic 67
- aorta abdominal normal 67
- aorta ascending arteriosclerotic 67
- aorta ascending normal 66
- aorta thoracic descending fibrous-arteriosclerotic 61, 66
- aorta thoracic descending lipid-arteriosclerotic 61, 66
- aorta thoracic descending normal 61, 66
- coronary artery arteriosclerotic 68
- coronary artery normal 61, 67
- pulmonary artery normal 61, 67
- vena cava inferior 61, 68
RNA polymerase 59

Thioctic acid, see Lipoic acid

Ubiquinone 76–84
- analytical procedure 76–79
- biological function 76
- in animal vascular tissue 79, 80
- in human vascular tissue 76, 79–83
Uridine diphosphate (UDP)
- bovine, aorta thoracic 64
Uridine diphosphate glucuronic acid (UDP-gluc.)

– bovine, aorta 58
Uridine monophosphate (UMP)
– bovine, aorta thoracic 58, 64
Uridine triphosphate (UTP)
– bovine, aorta abdominal 63
– – aorta thoracic 58, 64
– – carotid 63
– – coronary artery 63

Variation with age in coenzyme contents of animal vascular tissue 27, 28, 59, 61, 62, 64, 70, 71
Variation with age in coenzyme contents of human vascular tissue, *see* Coefficients of correlation between age and coenzyme contents